SpringerBriefs in Ethics

For further volumes:
http://www.springer.com/series/10184

Nathan Emmerich

Medical Ethics Education: An Interdisciplinary and Social Theoretical Perspective

 Springer

Nathan Emmerich
Visiting Research Fellow, School of Politics,
  International Studies and Philosophy
Queen's University of Belfast
Belfast
UK

ISSN 2211-8101   ISSN 2211-811X (electronic)
ISBN 978-3-319-00484-6 ISBN 978-3-319-00485-3 (eBook)
DOI 10.1007/978-3-319-00485-3
Springer Cham Heidelberg New York Dordrecht London

Library of Congress Control Number: 2013939342

© The Author(s) 2013
This work is subject to copyright. All rights are reserved by the Publisher, whether the whole or part of
the material is concerned, specifically the rights of translation, reprinting, reuse of illustrations,
recitation, broadcasting, reproduction on microfilms or in any other physical way, and transmission or
information storage and retrieval, electronic adaptation, computer software, or by similar or dissimilar
methodology now known or hereafter developed. Exempted from this legal reservation are brief
excerpts in connection with reviews or scholarly analysis or material supplied specifically for the
purpose of being entered and executed on a computer system, for exclusive use by the purchaser of the
work. Duplication of this publication or parts thereof is permitted only under the provisions of
the Copyright Law of the Publisher's location, in its current version, and permission for use must
always be obtained from Springer. Permissions for use may be obtained through RightsLink at the
Copyright Clearance Center. Violations are liable to prosecution under the respective Copyright Law.
The use of general descriptive names, registered names, trademarks, service marks, etc. in this
publication does not imply, even in the absence of a specific statement, that such names are exempt
from the relevant protective laws and regulations and therefore free for general use.
While the advice and information in this book are believed to be true and accurate at the date of
publication, neither the authors nor the editors nor the publisher can accept any legal responsibility for
any errors or omissions that may be made. The publisher makes no warranty, express or implied, with
respect to the material contained herein.

Printed on acid-free paper

Springer is part of Springer Science+Business Media (www.springer.com)

For Sally

Preface

This book is an attempt to think through the teaching and learning of medical ethics on the undergraduate medical degree. It would be easy to assume that ethics became a part of modern medical education as the result of external factors, notably bioethics. However, whilst it is certainly related to these wider developments, it is important that we see such education in its native context and relate the medical student's ethical education to their wider medical 'apprenticeship' and moral socialisation. The drive towards integrating medical ethics education across the curriculum demonstrates that whilst ethics can be seen as somewhat external to medical practice there is a recognition that it is important to engender a close relationship between theory and practice. Furthermore, despite Jonsen identifying a deeply meliorist orientation in the Birth of (American) Bioethics (1998:390), there has been limited bioethical research on how, and to what effect, medical ethics is taught to nascent medical professionals. Although there is some work in the tradition of 'moral education' we might contrast such concerns with the narrower focus on formal ethics taken here. What research there has been tended to focus on the psychological development of medical students. In addition little notice has been taken of wider research into the pedagogic process, of education, teaching and learning. Finally, there is a tendency in bioethics to mistake medical (or other professional) students for philosophy students and philosophical applied ethics for the moral and ethical practices embedded within modern medical culture. This book adopts a broader social theoretical perspective in an attempt to rectify some of our mistaken assumptions about what it is to teach medical ethics in a professional context.

To this end Chap. 1 introduces a number of concepts that frame and recur throughout the rest of the book. Of particular importance is the distinction, or lack thereof, between formal and informal education. Recent research has shown that the learning process does not differ whether or not it takes place in a formal classroom context of in an informal workplace context. This has obvious implications for medical education that occurs both at the bedside and in the classroom. Whilst the pedagogical content and the nature of what is learnt differ markedly across these two contexts the social psychology of the learning that takes place remains the same. No less important than the formal/informal distinction is the two metaphors for learning—acquisition and participation—identified by Sfard (1998).

Our standard assumptions about education are predicated on knowledge acquisition and its transfer from the teacher to the learner. Educational research in a variety of domains has shown the paucity of this model and encourages us to think about how knowledge is used and put into practice. During their education learners are inducted into this use. Therefore in order to capture this more active sense and as a compliment to the concept of socialisation I offer the notion of enculturation, an idea that is central to the rest of the work.

In Chap. 2 I discuss the social theory of Pierre Bourdieu and some recent sociological research into medical education that has made use of his analytic tools. I argue that Bourdieu offers a powerful perspective that allows us to understand and explore various aspects of what sociology calls professional reproduction. However, sociological research has a tendency to focus on the informal aspects of medical education and so, alongside some other commentators, I suggest that Bourdieuan social theory (and sociological perspectives on professional reproduction more generally) requires further development if it is to take proper account of the pedagogic process in its entirety. To this end, I discuss Bourdieu's concept of habitus in a cognitive frame and offer the concept of thinking dispositions as a tool that can underpin our understanding of ethics in professional practice and medical ethics education.

In Chaps. 3 and 4 I put aside these theoretical concerns and explore the recent history of medical ethics education in the UK and in the Medical School at Queen's University Belfast. There has been some research in this area that, in general, has focused on the London Medical Group, the precursor of the UK's Institute of Medical Ethics. I draw on this research and demonstrate that the activities and ends of this group were complimented by other changes in medical education that were taking place around this time. Particularly notable is the advent of general practice education and, in theories of professional practice and in approaches to professional education, the concept of reflection as well as a broader concern for the characterological and student's (professional) development. These connections can be clearly perceived in the career of WG Irwin who was the first Professor of General Practice at Queen's University Belfast, holding the position from 1971 to 1990. In developing general practice as an academic endeavour Irwin pioneered a number of educational innovations and, in collaboration with others including a 'consultant educationalist,' published a number of articles documenting his activities. Following his appointment to the Warnock Committee and having already taught terminal care he developed a concern for medical ethics. Whilst his medical ethics is idiosyncratic—being formed by seven somewhat overlapping principles—his pedagogic approach, developed in isolation from the kind of activities that characterised the London Medical Group and took place at other UK medical schools, is not overly distinct from the teaching that was developed in these institutions.

In Chap. 5 I return to more theoretical concerns and explore the recent rethinking of apprenticeship that has emerged from cognitive anthropology. Complimenting Bourdieu's social theory in its focus on practice and underpinned by a Vygotskian and therefore fully social psychological perspective I examine what this can bring to our understanding of medical ethics education. In the final analysis I consider what is actually enculturated as a result of medical ethics education. I suggest that this includes the concepts, style of reasoning (metacognition), and beliefs of modern medical ethics are legitimately considered as enculturated. However the way in which thinking dispositions are reproduced is better understood as a result of a cognitive socialisation that attends the enculturation of concepts, metacognition and beliefs.

In the conclusion I draw together the themes presented in the previous chapters. I suggest the views I have presented are bound together by an underlying dispositionalism. Adopting this perspective assists in countering some of the challenges to our ethical reasoning that have recently emerged from the new synthesis in moral psychology. I argue that we should perceive medical ethics as an orthogonal specialism of modern medical practice. Finally I draw out some of the implications for medical ethics education and, reflexively, for the practice of applied bio-ethics.

References

Jonsen, A.R. 1998. *The Birth of bioethics*. New York: Oxford University Press

Sfard, A. 1998. On two metaphors for learning and the dangers of choosing just one. *Educational Researcher* 27(2):4–13

Acknowledgments

This book is the result of my Ph.D. and I would like to thank my first, second, clinical and honorary supervisors, respectively, Prof. Lindsay Prior, Dr. Matt Wood, Dr. Maeve Rae and Dr. Ciaran Burke. I would also like to thank my external examiner, Prof. Alan Cribb. I would also like to make particular mention of Professor W. G. Irwin, who kindly corresponded with me and read an early draft of the chapter on his career, and Prof. Tony Hope, who corresponded with me about the Oxford Practice Skills Project.

Thanks are also due to my parents, particularly for their biannual contribution to my library.

Finally, I would like to thank Sally Wheeler, without whom this book, and the preceeding thesis, would never have been started, let alone finished.

Contents

Abbreviations

ASME	Association of Medical Education
BMA	British Medical Association
BMJ	British Medical Journal
DHSSNI	Department of Health and Social Services Northern Ireland
GMC	General Medical Council
GP	General Practitioner
IME	Institute of Medical Ethics
JME	Journal of Medical Ethics
LMG	London Medical Group
PBL	Problem-Based Learning
QUB	Queen's University Belfast
SCM	Student Christian Movement
SLT	Situated (or Socio-cultural) Learning Theory
SSME	Society for the Study of Medical Education
SSM	Student Selected Modules

Chapter 1
Some Relevant Concepts

1.1 Introduction

This book is an attempt to illuminate the educational activities of medicine in relation to ethics by pulling together a number of perspectives from a variety of disciplines, approaches and research programs. My specific concern is with what we are doing when we teach ethics to medical students and healthcare professionals. How it is that individual students and professionals learn about ethics, particularly as a member of (or as an part of becoming a member of) a group? This perspective includes considering medical students within particular classes or cohorts of other such individuals as well as in the wider context of the profession as a whole. Medical education aims at producing individuals who are well equipped and suitably qualified to join the medical profession and therefore to practice medicine. Medical ethics education aims to produce medical students who can engage in the *practice* of medical ethics (as distinct from simply putting medical ethics into practice). In order to facilitate the discussions that occur in subsequent chapters I should first like to introduce a number of concepts that together indicate the kind of perspective offered in this book. These concepts should give the reader an idea of how medical ethics education is approached, presented and therefore understood, in the context of interdisciplinary social theory. In short I try to make clear what it is to adopt a social perspective on medical ethics education and the connection this formal learning has with the medical apprenticeship.

1.2 Formal and Informal Education

The first distinction that we can draw is between 'formal' and 'informal' conceptions of education. Put simply: formal education occurs within a classroom setting whilst informal education occurs beyond its walls. This is not a perfect description, of course, as much formal learning takes place in, for example,

N. Emmerich, *Medical Ethics Education: An Interdisciplinary and Social Theoretical Perspective*, SpringerBriefs in Ethics, DOI: 10.1007/978-3-319-00485-3_1, © The Author(s) 2013

libraries. Informal education can also occur within the classroom walls. Formal pedagogic activities often, and unavoidably, communicate informal, often 'tacit' or 'hidden', information about, for example, the role of the Doctor and how it relates to other roles such as the nurse, the patient or the applied (bio)ethicist. It is certainly the case that not all of the teaching and learning that takes place under the aegis of the medical school is properly considered 'formal'. Furthermore the final years of a medical degree are predominately characterised by informal education as the kind of 'apprenticeship' learning that occurs during clinical placement is, essentially, of this type. One might then expect that, particularly in the context of attempts to 'integrate' medical education, there is not an absolute distinction to be made between formal and informal education but rather it is a spectrum and, simply, a useful way of characterising different contexts in which student are taught and learn. Certainly this has been the trend in the field of education since Scribner and Cole (1973) discussed the utility of the distinction. More recently Lave has suggested that whilst the distinction is assumed to map onto "so called context free and context embedded learning, or logical and intuitive understanding" (Lave 1997, p. 18) it exhibits a far greater complexity. Strauss (1984, p. 111) indicates some of this complexity in her arrangement of 'intentional' and 'incidental' learning, 'well-defined' and 'ill-defined' pedagogic procedures and some specific strategies (directing attention, rehearsal, chanting, other). Lave's research, discussed further in Chap. 5, rejects the view that the formal classroom is context free or 'a-cultural'. Furthermore much of the discussion presented in this book is aimed at developing the view that there is a connection between a reflective, reasoned, and enculturated understanding of medical ethics and a more intuitive, emotional and socialised appreciation of medical morality. In addition medical students develop both in relation to one another. Consequentially whilst we focus on formal and classroom based education in medical ethics there is an implicit commitment to the understanding of learning advanced by the sociology of medical education. This understanding is articulated in general terms of professional reproduction and medical socialisation. This latter is discussed below whilst a broader discussion of such research is offered in the next chapter.

However it is important at this juncture to demonstrate that the medical student's formal medical ethics classroom is not only embedded in the culture of modern biomedicine and medical education but is also an intrinsic aspect of that culture. In addition to modern medical ethics being characteristic of modern medicine per se it is the case that medical ethics education exhibits a professional approach to pedagogy. This can, to a degree, be found across the professions but it is also highly specific to medicine and also constitutive of its culture. There is some tendency, perhaps largely historical, usually implicit, and predominantly held by professional philosophers or ethicists, to consider the teaching of applied ethics to medical students as, fundamentally, no different from the teaching of applied ethics to philosophy students. This maps onto the view that 'medical ethics' is merely ethics applied to medicine, the so-called engineering model of applied ethics rejected by Caplan (1980). Just as this engineering model of applied ethics is misguided so is the idea that the philosopher's applied ethics classroom does not greatly differ from that of the medical students. Their pedagogic purposes are essentially different. The philosophy classroom is aimed the development of the philosophical, logical and reasoning skills of the students as ends in themselves. In contrast the medical student's classroom aims to develop these skills in order to facilitate ethical medical practice. Regardless of the *applied* nature of applied ethics, philosophy students are predominately engaged in an intellectual exercise that may or may not have practical implication. Medical students are, however, engaged in a *practical* exercise that is, in theory at least, designed to have a direct impact on their future professional activities. Of course a purely philosophical analysis may have a practical result or impact—many a philosophy student has embraced vegetarianism after reflecting on Singer's animal liberation (Singer 1995)—and it is certainly the case that less practical and more intellectual or abstract discussions can serve to inform the views of medical students in ethical matters and, in so doing, contribute to their moral and intellectual development, as professionals. Nevertheless, as revealed in the latter part of the preceding sentence, the function of an ethics education in a professional context is fundamentally different to that which occurs in a philosophical context.[1]

Demonstrating that there is a variation in purpose between the (applied) ethics classrooms of philosophy and medicine goes some way to demonstrating the presence of culture within formal classrooms. The point can be furthered if we take seriously the idea that medical ethics education was, at least in part, initially introduced as part of an attempt to change the moral and ethical culture of medicine. The motivation for professional medical ethics was not simply to address specific 'dilemmas' such as abortion, neonatal care or the availability of dialysis but also to engage with the changing nature of the Doctor-Patient relationship.

---

[1] This dichotomization of medical students and philosophy students education can, of course, be unsettled. For example medical students who intercalate in Healthcare Ethics, or similar, likely have less of an instrumental orientation to the content of these degrees, as do the tutors of these course. This is likely to be increasingly the case in the context of Masters level qualifications and for medical and healthcare professionals who undertake Doctorates, whether traditional, by publication or 'professional.' However, as this latter class of Doctorates reminds us, such qualifications and endeavours remain located within ideals of professional development and whilst individuals may have more or less instrumental motivations this remains the case as, ultimately, they return to practice newly informed by their intellectual exercises. I would argue that this remains the case for those who never return to practice and remain within their newly adopted discipline. Such individuals are not lost to medicine, healthcare or the professions but, rather, relate and contribute to it in a different way. If we take a wider view of the kind articulated here then we can see that these individuals are likely to occupy key positions in the social, cultural and intellectual relationship between medicine and other disciplines and, we might say, between the art and the science of medicine. Their contributions are, therefore, also likely to be key. Thus any unsettling of the view that the medical ethics education of medical students is more instrumental that that of philosophy students is then likely to reveal the unsurprising fact that the key change promoted by applied philosophical (bio)ethics in seeking to relate philosophy to medicine is that philosophy has a more instrumental and culturally embedded purpose and not that medicine, and its ethics, becomes a more abstract and intellectual endeavour.

In particular we might focus on the evolving need for a great level of interpersonal engagement and communication with patients. Further evidence for this point of view is offered in . It is perhaps sufficient for our current purposes to suggest that the introduction of ethics into the medical curriculum was related to a wider concern within medical education for the 'characterological.' Indeed within virtue ethics there is often a direct attempt to consider the characteristics (or virtues) that constitute the 'good' or 'ideal' doctor. It is self evident that this ideal varies across historical and contemporary cultural contexts. For example the ideal medical professional of the 1890s was very different to that of the 1990s and it is likely to be different again to that of the 2090s.

Here we might also think of professional role models. This is a central aspect of medical education, with particular relevance to the socialisation of medical students. Medical educators, especially those who work in the clinical context, are often individuals which, consciously or not, students model their future medical practice. This can, I think, be extended to medical ethics reasoning both in style and content or conclusion. As formal medical ethics education, particularly during the latter years of medical education, is often a cooperative undertaking which occurs between predominately philosophically orientated ethicists, clinicians and medical students we can conceive there to be a cultural and disciplinary dialogue taking place between medicine and bioethics.[2] Here the more practical concerns of medicine meet the more abstract approaches of 'pure' applied ethics. Post the birth of bioethics (Jonsen 1998) medicine, medical culture and, in particularly, medical ethics can no longer be considered the autonomous affair of the medical profession but neither can we consider bioethics to be an a-cultural phenomena or, indeed, the sole arbiter of medical morality (Fox and Swazey 1984). Rather bioethics has changed and developed medical culture and the formal medical ethics classroom is one example of this culture, or these cultures, at work. As an applied philosophical ethics bioethics has changed the way in which medicine *does* ethics. It clearly exhibits a particular cognitive style that often naturally leads to specific conclusions given its ontological and epistemic mode. Ethics, codified or not, has always proved to be a central aspect of a profession and provided grounds for its sociological status and cohesion (Freidson 1970, 2001; Macdonald 1995). Formal medical ethics education is therefore an essential aspect of the contemporary culture of the medical profession. Nevertheless, it remains only one aspect of modern medical morality. As such it will necessarily present informal, tacit and hidden lessons to medical students alongside and ingrained in its more formal and explicit pedagogical content.

---

[2] Veatch (2004) suggests this dialogue has only recently recommenced having been disrupted in the period 1770–1980.

1.3 Acquisition and Participation

Sfard (1998) has identified two metaphors—acquisition and participation—that construct and constrain our thinking with regard to learning and, therefore, education. Acquisition is, overwhelmingly, the predominant metaphor of learning found within our society. Here we find the mind construed as a kind of container for knowledge and education is, simplistically, the process through which knowledge is 'transferred' or 'decanted' from the minds of teachers and into the minds of students. In contrast, understanding learning to be the result of participation emphasises the embodied, active and temporally extended nature of learning. In a 'knowledge society' or 'economy' the idea of learning as the acquisition of facts has a prima facie appeal. This is furthered in the context of the medical profession, which is held to be predicated on a scientific i.e. objectively grounded, body of technical knowledge, in this instance the medical sciences. It is the acquisition of this knowledge that enables the professional to practice and so the aim of a professional education is, centrally, to pass on this knowledge to medical students. However, as mentioned, the later part of undergraduate medical education more obviously exhibits an apprenticeship format. As the alternative metaphor of participation has, in part, been developed as a result of research on learning in informal contexts we might expect it to offer insight into such professional pedagogic practices. Furthermore the predominance with which learning has been assumed to be associated with the supposed context-free formal educational setting has, one might suggest, a lot to do with the prevalence of acquisition as a metaphor for learning. Sfard has noted that were it not for the emergence of an alternative metaphor—participation- it is likely that we would not appreciate the way in which our educational discourses are if fact structured by 'acquisitionist' thinking, often to an overwhelming degree (Sfard 1998, pp. 5–6). Since this research programme has, reflexively, developed an appreciation for the informal learning that occurs within more formal contexts such as classrooms we might also expect that highlighting the ways in which medical students participate in the formal context to open up fruitful lines of enquiry.

Whilst the acquisitionist perspective belies the inarticualable knowledge that medical students require in order to engage successfully in future professional practice the participationist perspective can highlight this aspect of medical education. Perhaps the most obvious example of such knowledge is the tacit or embodied skills of the surgeon. It is impossible to instruct someone fully in the competent use of a scalpel using only formal means. The correct amount of pressure that one needs to apply to slice through skin, fat, muscle or cartilage is something that can only be acquired through experience or, to put it another way, through participation. Furthermore, as Schön (1984) demonstrated, the epistemological foundation of professional practice is not one of 'technical rationality' as would be implied by construing it as merely the rational application of some acquired body of knowledge. Rather it is the case that professionals exhibit knowing-in-action, something that involves a range of embodied and tacit abilities.

Thus the diagnosis of a patient does not take place via a computational process akin to a software program consulting a database with a list of symptoms. Rather it involves an intuitive grasp of scientific medical knowledge and how it relates to patients and patient complaints. Certainly formal reasoning and more obvious conscious reflections are involved in this process, particularly in complex situations. However this merely reinforces the truth that advanced, complex and expert practices are predicated on the relegation of standard, everyday and 'beginner' practices to the realm of the automatic and intuitive. In medical practice, as elsewhere, one must be able to walk before one can run. And one can do neither solely on the basis of acquiring the relevant intellectual knowledge.

The move to participatory perspectives on learning and education is part of what Roth has called a "paradigm shift... in thinking about thinking" (Roth 2001, p. 6). This involves moving beyond the computational metaphor of a mind which merely 'processes' information or independent data, to presenting mind, cognition and knowledge as inter-related phenomena. As a result the metaphor of participation "defies the traditional distinction between cognition and affect, brings social factors to the fore, and thus deals with an incomparably wider range of possibly relevant aspects" (Sfard 1998, p. 11) of learning and education. This approach construes mind as a historically and culturally situated phenomena something which is clearly an attractive proposition for both educators in general and for bioethics understood to be a cultural phenomena characterised be a meliorist ethos (Jonsen 1998, p. 392). Furthermore this perspective indicates that what is acquired in the course of medical ethics education cannot be achieved by a formal provision and presentation of an array of abstract concepts, principles or rules but rather implies the development of an "ability to communicate in the language of [the medical] community and act according to its particular norms" (Sfard 1998, p. 6). The medical communities moral or ethical activities are not restricted to merely doing the right thing but also being able to justify the right thing and to do so in the right way. Medical ethics education teaches medical students the concepts and skills they require for professional practice and it is only through participating in their use that students can properly be considered as acquiring them.

We might, at this juncture, note that "[w]hen speaking about human development, participationists do not mean a transformation in people, but rather in forms of human doing" (Sfard 2010, p. 80). This view of development is consistent with Rogoff's suggestion that it should be "assumed to proceed throughout the life span, with individuals' ways of thinking reorganising with successive advances in reaching and contributing to the understanding, skills and perspectives of their community. Examples.... include... retirement or parenthood; intellectual challenges; career changes; shifts in perspectives [regarding] patterns of relationships; [and] leaps of understanding of social institutions and interpersonal relationships" (Rogoff 1992, p. 11 emphasis added). Of particular importance is the proposed link between (social) psychological development and the reorganisation of ways of thinking. Medical students arrive at medical school as able as any other student to 'do' morality and ethics. However as part of becoming a medical doctor they must take on and develop a professional outlook which includes, indeed one might say is fundamentally constituted by (Freidson 1970; Macdonald 1995, p. 167), a specifically medical morality which incorporates professional medical ethics as a way, or mode, of thinking, talking and reasoning about ethics.

Sfard suggests that the expanded and developmental image of learning which emerges from the participationist view is one where "the permanence of having gives way to the constant flux of doing" (Sfard 1998, p. 6). This seems to me an attractive proposition as it indicates that the ethical education of medical students does not complete their professional moral development. If we see ethics as merely a matter of rational abilities then an ethical education is completed when the appropriate concepts and cognitive skills have been acquired. If we see it as the development of more than rational abilities but also of communicative and interpersonal skills, the ability to integrate appropriately ethical reasoning into the on-going practices of healthcare rather than merely 'put' ethics 'into' medical practice. This view has the potential to open up the possibility of a more nuanced understanding of the ethical practices of healthcare as well as for more diverse ideas, such as professional ethical leadership.

1.4 Socialisation and Enculturation

If we consider role models and role modelling as entailing aspects of leadership then the idea of *ethical* leadership in medical practice and education flows into a discussion of medical student's 'socialisation' and 'enculturation' as aspects of their medical education. The idea of socialisation has been widely used in sociologically inclined studies of medical education whilst the idea of enculturation is something I have developed (Emmerich 2011a) as a particularly useful tool for understanding formal ethics education. This is in part due to the conceptual connection it has with the more widely researched process of medical socialisation. Adopting a focus on socialisation has allowed those interested in the sociology of medical education, and those interested in professional reproduction more generally, to maintain an appropriate methodological stance towards their object of study. Anderson and Sharrock (1983) have called this stance 'irony-as-methodology' and, as its name indicates, it represents the degree to which sociological studies attempt to undermine 'official' or 'common-sensical' accounts of events. Such studies focus on demonstrating hitherto unperceived connections and motivations that structure the social world. In the case of medical education this has meant the neglect of formal pedagogical practices, directed attempts at education, in favour of a focus on the informal messages that are interwoven into the field of medicine, including medical practice, the medical school and the medical curriculum. This research has resulted in a variety of ideas and perspectives on medical education regarding what medical students learn during their medical education over and above what is contained in medical textbooks. The most notable, and most relevant, of which is the idea of medicine's 'hidden curriculum'

(Hafferty and Franks 1994; Hafferty and Castellani 2009). With the term enculturation I seek to preserve sociological insights into the moral and medical socialisation of medical students whilst developing a theoretically complimentary
perspective that can accommodate more formal education practices.

On the translation of some of his work into English the sociologist and
anthropologist Pierre Bourdieu, whose social theory we shall be discussing in the
next chapter, rejected the term socialisation and instead suggested the phrase
'collective enterprise of inculcation' (1977, p. 17), something Jenkins considers to
be "socialisation by any other name" (2010, p. 92). It remains a matter of
scholarly debate regarding what, if anything, he wished to convey either by his
refusal of this standard sociological term or by the replacement he proffered. It
certainly seems that socialisation is a collective enterprise of inculcation and vice
versa and Bourdieu is, in some circles, famous for failing to use one word when a
greater number might do the same job as well as for using complex words where
simple ones would do. However, if he meant anything by a collective enterprise of
then it is likely that he meant it to be a broader concept than mere socialisation.

Given his social theory it is likely that, in the first instance, he would highlight
the active participation, although not necessarily conscious and reflective participation, of the 'individual' or the individual's habitus in the process of socialisation. In the second instance he might well highlight the conscious, aware and
directed nature of our participation in the social world that is not merely the basis
for socialisation but an aspect of the on-going enterprise of inculcation. One
example of this might be the way in which we, particularly as adults, deliberately
engage in formal and directed forms of education such as a specifically medical
education. It is certainly the case that such an education has hidden, non-cognitive,
affective and un- non- or semi- conscious, effects which are rightly analysed in
terms of socialisation but this does not deny that there are open, aware, cognitive
and conscious processes which are also important for the (re)production of medical
professionals. No one would, or indeed has, denied this, at least not in an open and
outright manner. However these two processes have for the most part been considered independently from one another, if not as fundamentally independent
phenomena. This is an unsatisfactory state of affairs. If medical students learn in a
variety of ways those ways must, at some level, be fundamentally connected
otherwise why would they all be correctly thought of as aspects of a single pedagogic process? At the very least we should be able to posit and explore some
connection between the various ways students learn and at best we might seek to
connect them in a manner which demonstrates their fundamental interrelation i.e.
that they are part of the same 'collective process of inculcation.'

The term that I prefer for exploring formal educational practices in which
students are actively and consciously engaged is enculturation. My use of the term
has its roots in the anthropology of Herskovits who considered socialisation as
unconscious cultural conditioning whilst enculturation as a more specific and
psychosocial process of cultural transmission and transmutation (Shimahara 1970,
p. 144). He also distinguished between enculturation that takes place in the early
part of an individuals life—where it is implicated in the initial production and

stabilisation of wider culture norms in the minds of developing children—and in later life—where it is important for further, potentially directed, changes and developments in the previously acquired cultural repositories of individuals. Shimahara suggests that in adult life "an enculturative process... involves deliberations and choice amongst alternatives... [it] operates on a conscious level... [and] involves conscious reconditioning and reorientation" (Shimahara 1970, pp. 144–145).

This brief characterisation is enough to suggest that enculturation is the kind of concept that might compliment that of socialisation and facilitate the forging of a connection between informal and formal learning in medical ethics education. Furthermore we might bear in mind the fact that medical students are not unable to do 'ethics' or 'morality' prior to their arrival at medical school, as is the case with diagnosing Scarlet Fever or inserting a catheter. Rather it is the case that they have little sense of the moral landscape of medicine and cannot do *professional* ethics i.e. they are not yet attuned to the (moral) culture of medicine and cannot yet 'do' ethics in the prescribed way. Medical students can do ethics but they cannot do a culturally specifically *medical* ethics. Through engaging in reflection on the moral aspects of medicine (to which they are exposed) and with reference to a variety of concepts or principles (autonomy, beneficence, non-maleficence and justice, say) their pre-existing moral or ethical senses which, simplistically put, are the product of their childhood socialisation, are engaged, reconditioned and reoriented towards and into the professional context of medical practice and culture.

Shimanara is suggesting that enculturation takes places as part of reflective processes and proposes that we attend to the difference between (cultural) acquiring and (cultural) inquiring. She suggests "[e]nculturation has long been regarded as cultural acquiring" (Shimahara 1970, p. 148) and so, like socialisation, a passive process. However Herskovit's analysis reveals enculturation to be an active process of inquiry which results from the individual's *participation* in this active process with others and/or in a cultural context. Formal medical ethics education is one such active process of inquiry and so through their participation in this endeavour medical students are enculturated into medicine's culture of ethics. This entails the acquisition of a set of (medical) ethical concepts and an associated 'way of thinking' about the ethical issues of medical practice on the part of medical students. Having so acquired the culturally appropriate perspective and its terminology they may then begin to contribute to and participate fully in the ethical discourse(s) of medical practice. Coupled with our understanding of enculturation the on-going participation of medical professionals in ethical deliberation opens up the possibility for their on-going ethical development. No one would maintain that the ethical education of medical students ceases at the conclusion of their formal (undergraduate or postgraduate) education. Certainly, in the age of Continuing Professional Development (CPD), maintaining that the end of formal education signals the end of the need for continued learning in the arena(s) of biomedicine or professional medical practice is inadequate. The possibility for on-going ethical development and learning presents opportunity for a space in which maturity and leadership in the ethical aspects of medicine can be recognised. There is, I think, no possibility for such perspective in traditional, philosophical and analytic understandings of applied medical ethics and so it is a distinct advantage of this perspective that such a view can find purchase. Nevertheless there remains a need for a proper account of both (medical) ethical maturity and leadership if the concepts are to be taken seriously. Such an account may merely mitigate the possibility of adopting overly conservative views but at least this danger could be addressed. Regardless at this point we can, under the umbrella of a collective process of inculcation, accept the concept of enculturation as having a prima facie validity in regards medical ethics education and as a compliment to that of socialisation.

1.5 On Reflection

The idea of reflection has a wide degree of validity across the medical curriculum and the implications of the concept for professional practice and education are profound. However its relevance has not, in my view, been fully appreciated in the context of medical ethics education or by those medical ethicists who seek to contribute to medical education. In its current usage the term originates with the seminal work of Donald Schön (1984, 1990) whose research on a variety of occupations resulted in the rejection of 'technical rationality' as a basis for professional practice. The idea of technical rationality implies that professionals are involved in the application abstract, objective and value neutral knowledge (often scientific knowledge) to the problem(s) they wish to address. As noted above this same picture has been rejected in applied ethics under the term the 'engineering model' (Caplan 1980). Nevertheless there remains a widespread assumption that technical rationality is the basis of medical practice and that it adequately characterises the nature of applied ethics. Such assumptions regarding both of these cases can be found in lay people (patients), politicians, medical professionals, applied (bio)ethicists and moral philosophers. It can even be found in criticisms of these latter two groups, and bioethics more generally, where there often exists a 'straw man' concept of the discipline. Perhaps most importantly it appears to be a widespread assumption amongst medical students who must become reconciled to the challenges of uncertainly in medical practice (Fox 1957). In the case of ethics this 'uncertainty' is redoubled as there often appears a number of valid ethical perspectives to adopt towards a single case of concern.

In place of technical rationality Schön offers an epistemology of practice premised on 'knowing-in-action'. In part this deepens Ryle's distinction between 'knowing that' and 'knowing how' and Polyani's concept of 'tacit knowledge' (Schön 1984, pp. 51–52) through articulating the perspective that, first, practice involves forms of knowing that are inarticulable or uncodifiable and, second, that formal knowledge is dynamically embodied in practice. In the latter case we might again think of the physical abilities of surgeons that are irreducible to the formal knowledge that underpins surgical practice. Schön's view of knowledge as dynamically embodied in practice is represented by his view of knowing-in-action. As with the tacit knowledge of the surgeon's motor skills we might also think of their ability to work as part of a surgical team, within the particular local cultural arrangements (on the necessity for contextual acclimatisation when traversing medical cultures see: Harris 2011). Medical students acquire such knowledge through what Lave and Wenger (1991) call legitimate peripheral participation, a view of apprenticeship addressed in Chap. 5. The participation of such apprentices become decreasingly peripheral or, to put it another way, increasingly central until they become autonomous memvbers of the profession.

The rules that tacitly govern practice are not necessarily explicitly realised or realisable. Consider, for example, the ability of native language speakers to communicate in a grammatically correct manner whilst remaining unaware of formal grammatical rules. The relation of such rules to practice is uncertain, as they are not '*a priori*' precursors of grammatically correct speaking as much as they are a posteriori distillates of grammatically correct speaking. Schön (1984, p. 61) suggests that practitioners reflect on their knowing-in-practice and in this way it may be changed and developed. Thus, as with the developing grammatical practices of children, we can perceive a feedback loop between knowing-in-action and reflection. The practices that constitute the medical profession are not merely acquired by medical students though participation but they—the practices, the students and the reflective practitioner—are developed.

Schön also divides reflection into reflection-in-action and reflection-on-action, or to give the latter its full phrasing reflection-on-reflection-in-action (Schön 1990, p. 1). Reflecting-in-action usually involves the articulation of intuitive understanding and knowledge that is contextually present and Schön has likened it to an on-going conversation with the situation. It can be used to refresh or re-orientate the activity as it proceeds and it is particularly important where that activity is coordinated and involves a number of individuals (as in surgery) or where the understanding and knowledge of one party has to be communicated to another (as in a Patient-Doctor consultation). Reflecting-on-action can be used to learn lessons from experience as well as forming plans prior to, or during, professional activity. Obviously prior to even the most routine of surgeries there will be the formation and review of a plan. In case reviews the events so far, the prognosis and the future possibilities are reviewed. Putting such plans into action is not a simple matter of deciding and enacting but a complex process of reflecting, planning, considering eventualities, beginning activity, monitoring progress, on-going assessment and further reflection. However it is the potential for learning that, in this context, is most relevant.

As a pedagogic tool reflection attempts involves the full range of cognitive and affective activities that we see in professional practice. However here the aim is for the student to develop their understanding of the theory and practice of medicine; both the art and the science. Greenwood has described reflection-on-action as a cognitive post-mortem (Greenwood 1993, 1998) and we might suggest that certain aspects of medical education could be likened to an extended cognitive post-mortem; the dissection of medicine and medical practice as a whole and in its constitutive parts in order to build up an understanding of the medical professionals institutions and corporeal, practices on the part of the medical student. However we might suppose that the understanding already possessed by medical students is insufficient preparation for conducting such a task. This is, however, a flawed perspective as such a dissection is aimed at the on-going development of understanding: it is a layered process where initial surface understandings are overthrown in favour of more complex and deeper appreciations. Furthermore we might recall that medical students are not merely being educated in medical practice but in *reflective* medical practice. Thus we might conclude that what is being developed is not only an understanding of medical practice but also the student's reflective abilities. There is then a strong sense in which the pedagogic medium of professional education, reflection, is also the pedagogic message: it is through their reflective education that students come to be capable of reflective practice.

This short characterisation of Schön's ideal, and we should remember it is an ideal, of reflective practice and education in the context of medicine is sufficient for our subsequent purpose as, at this juncture, I wish to delimit my talk of reflection by making use of the related conception of metacognition. There are a number of reasons for this, not least of which is the sheer range of literature which develops and (re)interprets reflection in the context of professional medical and allied health education. We might note the emotional aspect to Ghaye and Lillyman's account (Ghaye and Lillyman 2010) that has been entirely omitted from the discussion above but is vital to fully comprehending the utility of reflection in the context of medical education that has obvious emotionally challenging aspects (cf. Maxwell 2008). We might also consider a number of other theoretical varieties. Not only do we have accounts of reflection (Ghaye 2010) but also 'critical reflection' (Fook et al. 2006) and 'reflexive practice' (West 2010). Furthermore the success of the reflective paradigm has led some to proclaim its comprehensive nature. LaBoskey and Hamilton, for example, consider it to be "not *a* way, but *the* way for making meaning that is justified—empirically, theoretically, morally" (2010, p. 334 emphasis added). The risk here is that reflection becomes a term that names everything and explains nothing. Whilst remaining appreciative of the reflective turn in (professional) education I see no reason to add to this already substantial and diverse discourse particularly where the focus is on the specific practices of formal medical ethics education. Thus I offer metacognition as a specific strategy that permits a focus on formal medical ethical reflection and education as a way of thinking. Nevertheless it remains situated within the wider conception of (professional) reflection and reflective practice.

1.6 Metacognition

It is clear that within the literature on reflection in professional practice and education there is a concern with 'thinking.' Alongside this literature one can also find more focused discussion of 'thinking' in education (Lipman 2003; Kuhn

2005) that also makes use of the term 'reflection'. However whilst there is here a concern with teaching as a professional practice it is the case that rather than seeking to indicate something about the epistemic basis of professional practice, and so professional education, the term is used in a manner that is perhaps closer to what Socrates' might have been advising when he said that the "unexamined life is not worth living" (Plato 1997, § 38a). It is certain that the idea of reflection has its own meaning in discourses regarding 'thinking' in education as compared the 'reflective paradigm' in professional education. This is clear if we consider that in the former the focus is, for the most part, on compulsory education and not professional, or even higher, education. Nevertheless in this work there is a recognition that "the reflective model is thoroughly social and communal" (Lipman 2003, p. 25) and so it resonates with understandings of professional reflection as well as the participation metaphor of learning. It is consistent with that aspect of reflection that recognises that "thinking that is aware of its own assumptions and implications as well as being conscious of the reasons or evidence that support this or that conclusion" (Lipman 2003, p. 26) which is the focus in the following discussion. In order to avoid confusion with the wider concept of reflection I am going to call this metacognition and draw on the work of others, notable, Kuhn et al. (1988) who have also made use of the term, albeit for different initial motivations.

The concept of metacognition has made an appearance in medical education research, notably Quirk's 'Metacognition and Intuition' (2006). However this work adopts an individualist view unsuited to the more social view being presented here. Furthermore one could read Quirk as offering a rereading of 'reflective practice' that collapses and dichotomises the more sophisticated and nuanced understanding offered by Schön into 'metacognition' and 'intuition.' Thereby he does what Johannessen has criticised others for doing which is to "hide all the complexity behind the term intuition" (Johannessen 2006, p. 242). This renders metacognition, which Quirk considers a kind of intelligence (2006, p. 25), a much thinner concept than Schön's idea of reflection and one that is, implicitly, reliant on an epistemology of technical rationality. Albeit one that operates in concord with 'intuition' and which seeks to be self-critical or self-adjusting. Quirk's individualism also poses problems for a sophisticated view of student and professional development. Metacognition, it is supposed, is an 'innate' ability which is merely 'acclimatised' to medicine and medical culture. The view I wish to present is that the metacognitive abilities developed by medical students should not be considered independent of their understanding of medicine and medical practice. Metacognitive abilities do not merely become acclimatised to a new context but, rather, there is a mutual co-construction and development of the individual's abilities alongside their understanding of and appreciation for the context to which they relate. The ground on which I wish to articulate this claim is the formal medical ethics education of medical students which can now be construed as aiming to developing particular (meta)cognitive skills of medical students (teaching then to use ethical concepts to analyse medical ethical cases) as well as attempting to develop their understanding of medicine as an ethical enterprise (or, perhaps less ambitiously and more realistically in the early stages of medical students moral and ethical development, as an enterprise which exhibits a particular typography captured by the principles of medical ethics).

However prior to taking up this line of enquiry in the next section, and in the final chapter, a more detailed picture of metacognition is required. With Quirk I agree that medical students arrive at medical school with certain, and to a degree transferable, metacognitive abilities resulting from their previous education. However these abilities are not asocial or acultural, and nor are they 'innate' skills. Rather they are the complex product of the socio-culturally located individual and their social trajectory that, of course, includes their previous (formal and informal) educational experiences. Such individuals are, developmentally, products of their particular socio-cultural contexts; contexts which include a variety of 'learning communities'. One such community is what Mercer calls a thinking community (2000, pp. 170–171) whilst Lipman promotes a more ideal, or more generalised, vision of a Community of Inquiry (CoI) (2003, p. 20). In the context of medical education we can certainly think of formal educational practices in the terms offered by Mercer and Lipman whilst also acknowledging that this is part of a wider pedagogic process that includes informal education and more obvious examples of apprenticeship learning. We might then consider the 'thinking community' or CoI found within the formal medical ethics classroom as simply one aspect of a wider Community of Practice (Wenger 1999) that is the medical profession and includes medical education. Furthermore, as discussed further in Chap. 5, we might adopt a broader understanding of apprenticeship learning as involving 'legitimate peripheral participation' (Lave and Wenger 1991) in a community of practice. To do so is, once again, to highlight the culturally situated nature of formal educational practices and the essential connection between the formal and informal contexts of learning in medical education.

This focus on community can be taken as one indication of the importance of language and dialogue for the development of thinking skills, such as metacognition, particularly where they are predicated on the use of certain conceptual apparatus such as the principles of medical ethics. We come to understand concepts not through formal, ostensive or verbal, definitions but through making use of them (Miller and Gildea 1987; Brown et al. 1989). Furthermore we make use of them within particular cultural contexts and within what Wittgenstein would call language games (2009, § 23). Certainly the ability to give and understand verbal definitions is itself an important language game and, furthermore, such definitions can be considered particularly relevant to contemporary casuistic and case-based (applied) medical ethics. Nevertheless ostensive definitions of words are necessarily insufficient to circumscribe or prescribe fully their use(s). Such definitions cannot contain all the rules for their own application. It is in this context that we should understand the introduction of medical ethical terms to medical students. Under this view we should not "understand language use simply in terms of information transfer between people... [rather it is] a collaborative endeavour in which meanings are negotiated and some common knowledge is mobilised... Language is designed for doing something... it allows the mental resources of individuals to combine in a collective, communicative intelligence which enables people to make better sense of the world and to devise practical ways of dealing with it" (Mercer 2000, p. 6).

Writing about the development of scientific thinking skills Kuhn et al. note that Vygotsky, whose psychological theory I discuss in Chap. 5, "regarded conscious awareness as the heart of metacognition" (1988, p. 32) and they highlight meta-cognition as "reflection on one's own thinking" (1988, p. 228). They make par-ticular reference to metacognition as "reflection on one's own theories—as thinking about a theory rather than with it" (1988, p. 32) but also suggest that it involves the "[s]kilful coordination of theory and evidence" (1988, p. 3) and particularly when there exists an "awareness and control over this interaction" (1988, p. 228). Metacognition is then more than merely "thinking about one's own and another's thinking and feeling" as Quirk suggests (1996, p. 24). As I have attempted to draw out in the previous paragraph "a single thinker cannot under-stand (hold concepts about) the world—cannot even hold any beliefs about it (or themselves) unless she is in a certain kind of interpretive or dialogical relationship with others" (Splitter 2011, p. 355). Elsewhere in her research Kuhn is clear that, in her view, dialogical reasoning precedes monological reasoning as this latter skill can only be acquired and developed by individuals through participation in dia-logical argumentation with their peers and others (Kuhn 2005, pp. 114–115). Therefore, as Splitter suggests, any "engagement in the quest for knowledge is equivalent to my own dialogical engagement with others" (Splitter 2011, p. 355). Medical ethics is a metacognitive ability and the quest for knowledge about the ethical dimensions of medicine conducted by medical students in the context of formal (and informal) medical education is therefore characterised by dialogical engagement with their peers, medical educators and, reflectively, themselves.

Whilst Kuhn, Amsel and O'Loughlin think of metacognition as "general skills, in the sense of being definable across a wide range of content" (Kuhn et al. 1988, p. 229) they "do not see the development of these thinking skills as deriving from an underlying logical competence" (Kuhn et al. 1988, pp. 228–229), a perspective we could impute to Quirk (2006). Kuhn et al. view of metacognition is one of a linguistically based cognitive ability which emerges "in pragmatic, goal-related contexts, in terms of which they are originally defined" (Kuhn et al. 1988, p. 229) and, we might add, in which they subsequently re-emerge and become redefined. Only through engaging in further metacognitive activity can it become dialogically reappropriated into new contexts. Metacognitive thinking skills are not something we individually acquire nor can they be simply applied to new, previously un-encountered, domains. Rather it is the case that metacognitive abilities are built up and developed in an on-going manner though engagement in dialogical meta-cognitive activities that occur in particular circumstances and contexts with par-ticular aims and purposes. Within this perspective we can then differentiate between the metacognitive abilities developed by philosophy students and medical students in their respective applied ethics classrooms.

This thought, that metacognitive abilities are relative to particular purposes and contexts, can also be applied to the domain of medical science and medical expertise (Patel et al. 2000). Just as the abstract scientific concepts and knowledge of biomedicine must be brought into the context of practice the abstract or formal teaching of some generalised principles or abstract concepts of medical ethics is not sufficient to equip medical students with the skills to make use of those concepts in the variety of circumstances that may present in medical practice. Rather medical students must be taught medical ethics and subsequently be encouraged to explore these concepts in a variety of formal and informal contexts, particularly in relation to a variety of specific cases, preferably ones which relate to their practical experiences. It is through this approach that generalised metacognitive abilities in the field of medical ethics are more successfully enculturated into medical students. As Kuhn, Amsel and O'Loughlin acknowledge, "metacognitive control of a strategy [is] acquired through exercise [which] promotes transfer and increases generality. Paradoxically, then, the exercise of strategies within very specific, content-delimited contexts may promote their generalisation, whist didactic teaching of the same strategies in more abstract, general form may fail to achieve this same end" (Kuhn et al. 1988, p. 231). It is only through engaging in particular instances of medical ethical reasoning in relation to specific cases that medical students will come to develop a broader understanding and appreciation for medical ethics as being underpinned by a coherent theoretical perspective. It is the understanding developed in the context of particular cases that supports an understanding of the theoretical generalisation, rather than the other way around. Metacognitive competence is achieved when both the particular case and the generalised conceptual perspective can be mutually coordinated by individual medical students or professionals in the context of their (community of) professional practice.

1.7 Learning to Talk, Learning to Think: The Dialogical Nature of Socio-Cultural Located Argumentation

The kind of position articulated above is consistent with the view that all propositional thinking is, fundamentally, a linguistic phenomena. Splitter, for example arguers that "[c]oncepts cannot be separated from the language *used* to define and identify them... they are as much 'dialogical' objects' as they are objects of our thought or introspection.... Higher levels of thinking [such as reflection and metacognition] *are* linguistic... [D]ialogue both reflects and generates what it constitutes, namely our very existence as thinkers, each of whom can flourish as only one-among-others." (Splitter 2011, p. 350) italics in original, underlining added). It is not the case that language is a precondition of conceptual thinking or that conceptual thinking is a precondition of language rather they are inseparably instantiated phenomena. Whilst it appears as if the interiority and individuality of 'thinking' or 'reflection' distinguishes it from the external and interpersonal nature of linguistic communication (whatever semiotic form it might take, written, spoken, signed etc.) the two are essentially intertwined. As Splitter continues "dialogue both reflects and generates what it constitutes, namely our very existence as thinkers, each of whom can flourish as only one-among-others" (Splitter 2011, p. 350). Therefore, as Kuhn (2005, pp. 114–115) has made clear, rather than being the basis or foundation of dialogue, monologue and monological abilities can only emerge as a result of dialogical abilities. Thus the 'single thinker,' the paradigm of Western, analytic, and scientific authority embodied in Rodin's sculpture, Descartes Meditations, and analytic philosophy—including applied (bio)ethics—more generally, can only emerge from within a community of such individuals.

As we have identified formal medical ethics education to, in essence, be involved in the transmission of a set of linguistic concepts to medical students which cannot be achieved outside of engaging in, participating in, their use, then we can see the kind of dialogue produced is essential to individual medical students becoming competent users of, which is to say competent reasoners with, the concepts. If it is the case that our reasoning works with and through sets of culturally defined and embedded conceptual tools then in reasoning and thinking with them we are making use of them in the same way as we do when using them to communicate. We can now see that language and linguistic concepts define and "organise categories of reality and structure ways of approaching situations… [and they can then] be regarded as technologies for organising thought process that are utilised in individual problem solving and practiced in institutional activities" (Rogoff 1992, p. 51). Melser suggests that given this view there is cause to perceive thinking itself as a kind of doing or act(ion) (2004, p. 4) by which he means that "[a]part from one or two unusual features thinking is an ordinary, learned and voluntary" (2004, p. 137). What makes thinking an act is our ability to voluntarily and in a self-aware manner direct our reflective attention. Furthermore we learn to do so and our ability to do so can be evaluated. It is often observed in facial expressions and bodily reactions (and we might add in our forms of communication) (Melser 2004, pp. 6–12). Thus whilst Melser does not emphasise or directly engage with the social and linguistic nature of thinking, as I have done in discussing metacognition, we may still appreciate his view of thinking as an activity and take it as being consistent with the more explicitly social perspective adopted here.

We might also conclude that, understood as an act or activity, thinking is a practice. Therefore educating for (or enculturating) such forms of thinking is, in Sfard and Rogoff's terms, developmental. Social-psychological accounts of development, as articulated by participationist perspectives, are predicated on changes in practice and 'ways of doing' (Sfard 2010, p. 80). Medical education (training, socialisation and enculturation) is characterised by such changes as medical student's traverse various stages that taken them from an initial inability to practice medicine in any way, shape or form, to a state of dependant (and interdependent) practice and, finally, to being a competent and independent practitioner. Medical ethics education should be understood as not only recapitulating this trajectory but as being an aspect of this process of individual and collective professional reproduction.

1.8 Conclusion

In this chapter I have attempted to provide an overview of some of the concepts and perspectives that inform the views presented in the remainder of the book. This chapter has also attempted to present the case for considering ways of talking' and 'thinking' as forms of practice. As the sociological perspectives of Freidson, MacDonald and many others have shown ethics, codified or not, are central to the constitution of a profession. Thus what it is to practice as a professional is always, explicitly or implicitly, an inherently moral matter and implies conformity to a certain ethical standard. The advent of modern medical ethics has rendered these ethical standards explicit and so the practice of medical professionals now includes ways of talking and thinking about ethical concerns such that they are addressed in an appropriately explicit manner. In this view the justification of the practice is inseparable from the practice itself. Medical ethics education does not aim to teach medical students a set of stock answers to the ethical problems of medicine which can then remain implicit in practice. Despite the fact that there are generically identifiable kinds of ethical problems or 'cases' that medical professionals regularly face and 'solve' or address in recognisably consistent ways, medical ethics education does not aim at simply socialising medical students into the standard ethical practices and responses of medical culture. Rather they are enculturated into the use of a particular language, a particular way (or mode) of (reflective and metacognitive) thinking, that enables the medical professional, and the medical profession, to engage in a practice that explicitly addresses the ethical dimensions of medical as an aspect and requirement of modern medical practice. Medical ethics is a second-order normative phenomenon. It is the 'way of thinking that is prescribed rather than, simply, the answers it produces.

References

Anderson, D., and W. Sharrock. 1983. Irony as a methodological theory: A sketch of four sociological variations. *Poetics Today* 4(3): 565–579.

Bourdieu, P. 1977. *Outline of a theory of practice*. Cambridge: Cambridge University Press.

Brown, J.S., A. Collins, and P. Duguid. 1989. Situated cognition and the culture of learning. *Educational Researcher* 18(1): 32–42.

Caplan, A. 1980. Ethical engineers need not apply: The state of applied ethics today. *Science, Technology and Human Values* 6(33): 24–32.

Emmerich, N. 2011a. *Taking Education Seriously: Developing Bourdieuan Social Theory in the Context of Teaching and Learning Medical Ethics in the UK Undergraduate Medical Degree*. Unpublished PhD Thesis, Queen's University Belfast.

Fook, J., S. White, and F. Gardner. 2006. Critical reflection: a review of contemporary literature and understandings. In *Critical reflection in health and social care*, eds. White, S., J. Fook, and F. Gardner, 3–20. Maidenhead: Open University Press.

Fox, R.C. 1957. Training for uncertainty. In *The student physician*, eds. Merton, R. G.G. Reader, and P.L Kendall, 207–241. Cambridge: Harvard University Press.

Fox, R.C., and J.P. Swazey. 1984. Medical morality is not bioethics–medical ethics in China and the United States. *Perspectives in Biological Medicine* 27(3): 336–360.

Freidson, E. 1970. *Profession of medicine. A study of the sociology of applied knowledge*. USA: University of Chicago Press.

Freidson, E. 2001. *Professionalism, the third logic: On the practice of knowledge*. USA: University of Chicago Press.

Ghaye, T. 2010. A reflective inquiry as participatory and appreciative action and reflection. In *Handbook of reflection and reflective inquiry*, ed. Lyons, N, 553–569. New York: Springer.

Ghaye, T., and S. Lillyman. 2010. *Reflection: Principles and practice for healthcare professionals*, 2nd ed. Great Britain: Quay Books.

Greenwood, J. 1993. Reflective practice: a critique of the work of Argyris and Schön. *Journal of Advanced Nursing* 18(8): 1183–1187.

Greenwood, J. 1998. The role of reflection in single and double loop learning. *Journal of Advanced Nursing* 27(5): 1048–1053.

Hafferty, F.W. and B. Castellani. 2009. The hidden curriculum: A theory of medical education. In *Handbook of the sociology of medical education*, eds. Brosnan, C and B.S. Turner, 51–68. Abingdon: Routledge.

Hafferty, F.W., and R. Franks. 1994. The hidden curriculum, ethics teaching, and the structure of medical education. *Academic Medicine* 69(11): 861–871.

Harris, A. 2011. In a moment of mismatch: overseas doctors' adjustments in new hospital environments. *Sociology of Health and Illness* 33(2): 308–320.

Jenkins, R. 2010. Pierre Bourdieu: From the model of reality to the reality of the model. In *Human agents and social structures*, eds. Martin, P.J. and A. Dennis, 86–99. Manchester: Manchester University Press.

Johannessen, K.S. 2006. Knowledge and reflective practice. In *Dialogue, skill and tacit knowledge*, eds. Göranzon, B., M. Hammaren. and R. Ennals. 229–242. Chichester: Wiley.

Jonsen, A.R. 1998. *The Birth of bioethics*. New York: Oxford University Press.

Kuhn, D. 2005. *Education for thinking*. Massachusetts: Harvard University Press.

Kuhn, D., E. Amsel, M. O'Loughlin, and H. Beilin. 1988. *The development of scientific thinking skills*. London: Academic Press.

LaBoskey, V.K. and M.L. Hamilton. 2010. Professional pedagogies and research practices: Teaching and researching reflective inquiry. In *Handbook of reflection and reflective inquiry: Mapping a way of knowing for professional reflective inquiry*, eds. Lyons, New York, USA: Springer N, 351–381.

Lave, J. 1997. The culture of acquisition and the practice of understanding. In *Situated cognition: Social, semiotic, and psychological perspectives*, eds. Kirshner, K. and J.A. Whitson, 17–36. Mahwah: Lawrence Erlbaum Associates.

Lave, J., and E. Wenger. 1991. *Situated learning: Legitimate peripheral participation*. USA: Cambridge University Press.

Lipman, M. 2003. *Thinking in education*, 2nd edition. Cambridge: Cambridge University Press.

Macdonald, K.M. 1995. *The sociology of the professions*. London: Sage.

Maxwell, B. 2008. *Professional ethics education: Studies in compassionate empathy*. Dodrecht: Springer.

Melser, D. 2004. *The act of thinking*. USA: MIT Press.

Mercer, N. 2000. *Words and minds: How we use language to think together*. Oxford: Routledge.

Miller, G.A., and P.M. Gildea. 1987. How children learn words. *Scientific American* 257(3): 94–99.

Patel, V.L., R. Glaser, and J.F. Arocha. 2000. Cognition and expertise: Acquisition of medical competence. *Clinical and Investigative Medicine* 23(4): 256–260.

Plato, 1997. Apology, project gutenberg. http://www.gutenberg.org/ebooks/1656. Accessed 1 July 2011.

Quirk, M.E. 1996. *How to learn and teach in medical school: A learner-centered approach*. Springfield: Charles C. Thomas Publisher.

Quirk, M.E. 2006. *Intuition and metacognition in medical education*. New York: Springer.

Rogoff, B. 1992. *Apprenticeship in thinking: Cognitive development in social context.* New York: Oxford University Press.

Roth, W.M. 2001. Enculturation: Acquisition of conceptual blind spots and epistemological prejudices. *British Educational Research Journal* 27(1): 5–27.

Schön, D.A. 1984. *The reflective practitioner: How professionals think in action.* USA: Basic Books.

Schön, D.A. 1990. *Educating the reflective practitioner: Toward a new design for teaching and learning.* San Francisco: Jossey Bass.

Scribner, S., and M. Cole. 1973. Cognitive consequences of formal and informal education. *Science* 182(4112): 553–559.

Sfard, A. 1998. On two metaphors for learning and the dangers of choosing just one. *Educational Researcher* 27(2): 4–13.

Sfard, A. 2010. *Thinking as communicating: Human development, the growth of discourses, and mathematizing.* New York: Cambridge University Press.

Shimahara, N. 1970. Enculturation—A reconsideration. *Current Anthropology* 11(2): 143–154.

Singer, P. 1995. *Animal liberation.* New York: Random House.

Splitter, L.J. 2011. Agency, thought, and language: Analytic philosophy goes to school. *Studies in Philosophy and Education* 30(4): 343–362.

Strauss, C. 1984. Beyond 'Formal' versus 'Informal' education: Uses of psychological theory in anthropological research. *Ethos* 12(3): 195–222.

Veatch, R.M. 2004. *Disrupted dialogue: Medical ethics and the collapse of physician/humanist communication, 1770–1980.* New York: Oxford University Press.

Wenger, E. 1999. *Communities of practice: Learning meaning and identity.* Cambridge: Cambridge University Press.

West, L. 2010. Really reflexive practice: auto/biographical research and struggles for a critical reflexivity. In *Bradbury, H*, ed. N. Frost, S. Kilminster, and M. Zukas, 66–80. New approaches to professional lifelong learning: Beyond Reflective Practice.

Chapter 2
Sociological Perspectives on Medical Education

2.1 Introduction

The inaugural editorial of the British Journal of Medical Education (now simply 'Medical Education') claimed 'medical education' as "one of the subjects of medicine" (1966, p. 1) and, therefore, a legitimate *medical* specialism. Norman (2011) identifies three distinct generations of medical education researchers it has, for the most part, been conducted by those involved with medical education itself and with a view to improving and developing the pedagogic practices that form the basis of their concern. Such research can be distinguished from the sociology of medical education. Whilst these two endeavours can, and perhaps ideally should, closely inform one another they are often sharply distinct. This is reflected in the fact that whilst medical education was present at the inauguration of medical sociology, in the shape of Becker et al.'s Boys in White (1961) and Merton's et al.'s Student Physician (1957), little was done to build on this foundation until recently (Jefferys and Elston 1989). Nevertheless the field of medical education research has grown steadily in the intervening decades.

The past 15 years have seen a resurgence of interest and research on medical education from a sociological perceptive. There have been a number of doctoral dissertations and associated publications Brosnan (2008, 2009, 2010; Lempp 2004; Luke 2003; Sinclair 1997) and a recent handbook (Brosnan and Turner 2009a). This chapter draws on this research. Given the trend towards configuring the university medical school as independent from biomedical research, the imperative to conduct research, and the prevalence of professional non-clinical medical educators, it is likely that medical education research will continue to increase. Whilst some have expressed concern about the theoretical sophistication of such research this should be understood as a call for the professionalisation of the discipline and for greater level of mutual engagement between 'theoretical' and 'applied' researchers (Albert and Reeves 2010).

Possibly because it offers a sophisticated perspective on social and professional reproduction Bourdieuan social theory is a notable feature of much recent sociological research into medical education. However Robbins (1993) identifies a

N. Emmerich, *Medical Ethics Education: An Interdisciplinary and Social Theoretical Perspective*, SpringerBriefs in Ethics, DOI: 10.1007/978-3-319-00485-3_2, © The Author(s) 2013

weakness in Bourdieuan theory suggesting it often neglects the formal cognitive content of educational activities. Thus the potential for this perspective to connect with research into medical education has not, yet, been fully realised. The influential article by Hafferty and Franks (1994) on the hidden curriculum is an example of how sociological and education perspectives can compliment one another there remains a need for further engagement. Bourdieu's social theory offers significant potential in this regard. As Harker has suggested his "theory of cultural practice develops the notion of reproduction to far higher levels than can be found in any… specifically educational writing. [Nevertheless] Bourdieu himself does not re-address educational issues from this advanced theoretical perspective" (1984, p. 125). This chapter attempts to develop Bourdieuan social theory in the manner required "to re-insert the consideration of course or cognitive content into the [Bourdieuan] analysis of the teaching and learning process" (Robbins 1993, p. 162). To this end I offer the concept of thinking dispositions as something produced by formal medical ethics education and as a conceptual underpinning for the development of a cognitive dimension to habitus.

The habitus is a central concept in Bourdieuan social theory. It is the embodied and habituated dispositions of individual's and the principle resource of their practical abilities. As such the focus for much of the sociological research into medical education is medical socialisation and directly concerned with the development of a professional medical habitus and the associated dispositions. In the next section I offer some further detail on Bourdieu's social theory before reviewing some of the recent, predominantly Bourdieuan, sociological research into medical education. Finally I situate the broadly psychological idea of thinking dispositions in a Bourdieuan context suggesting it can theoretically augment my perspective on the ethical enculturation that attends the development of medical habitus.

2.2 The Social Theory of Pierre Bourdieu

The social theory of Pierre Bourdieu is a theory of practice. Such theories attempt to overcome the dichotomy between structure and agency and so the invidious choice between sociological determinism and individual freewill. They seek to connect social structure and human agency through reconceiving society as the product of the ongoing discursive activity of individuals and groups. This emphasis reflects the metaphor of learning as participation (discussed in Chap. 1) where education is conceived as a discursive activity, or practice, engaged in by both teachers and students. Bourdieu perspective is particularly well suited to the theorisation of professional reproduction, the main aim of medical education. He emphasises the interaction of the field with the habitus, of the social structure with the individual, of the medical school with medical educators and, more importantly, medical students. Facilitated by habitus it is the practices of individuals that reproduce the field. In turn exposure to the field results in the 'collective enterprise of inculcation'—the socialization and enculturation of individual's habitus—and, therefore, the reproduction of practice(s).

As discussed in the previous chapter Bourdieu rejects the term socialisation preferring "the collective enterprise of inculcation" (Bourdieu 1977, p. 17). One motivation for this may have been because socialisation has (or had) a psychological, rather than sociological or anthropological, flavour that Bourdieu would not have found attractive. However his rephrasing also reflects the way in which individuals play a complicit and even active role in their own socialisation. To be socialised is not a passive experience but involves the participation of the individual in particular social contexts. In addition, whilst Bourdieu never really emphasised the cognitive aspect of habitus, practice and social reproduction, the phrase collective enterprise of inculcation can, I have argued, be understood as reflecting the idea of enculturation. I further develop this aspect of Bourdieuan social theory in the final section of this chapter on thinking dispositions.

Bourdieu's perspective is a sophisticated and extensively articulated theory that encompasses a range of powerful conceptual tools. Individual discussion of most of these concepts can be found in Grenfell (2008) and a comprehensive introduction to 'reflexive sociology' in a Bourdieuan frame can be found in Bourdieu and Waquant (1992). Of necessity my discussion is limited and I focus on the concepts of practice, habitus and field with reference to medical practice and education.

2.2.1 Practice, Field, Habitus

As discussed 'practice' is a concept conceived, at least in part, in order to overcome the dichotomy of the individual as an unfettered rational actor or as a determined product of social structures. Conceptions of practice reject the notion that human activity is the product of rules. Rather they are improvised on the basis of our embodied senses. Consider the activity of riding a bike. It is conceivable that we could come up with a set of rules for bike riding that would allow us to design and program a robot to perform this activity. However there is no sense in which human beings are following such rules when they ride bikes. Rather through practice (or 'rehearsal') we develop a sense of balance and an ability to coordinate activities such as peddling and steering that allows us to perform the task of cycling successfully. Furthermore whilst there might be more obvious and codified 'rules of the road,' a highway code that cyclists should follow when negotiating the public highways, these also become embodied. Certainly novice cyclist can learn the rules of the road by rote but the competent cyclist achieves this status through no longer needing to consciously think about the rules and being able to follow them unconsciously.

From the perspective of practice theory social life is sufficed with practices of this sort. Our ability to walk down a crowded street, hold a conversation, negotiate a market-place, create a home, or do our jobs are all practices based on tacit rules that we may have never consciously learnt and may not, ultimately, be fully expressible in a propositional sense. Rather we have been socialised into a set of social and cultural practices, some of which include stated 'rules' and some of which do not. In the context of medical practice we might think of surgery as analogous to riding a bike. Certainly surgery involves a good deal more knowledge than is given in the Highway Code but it is, nevertheless, an embodied skill acquired through extensive practice. Whilst the competent, indeed expert, surgeon will attend a theatre with a plan, or map, of the operation their performance relies on their embodied ability to use a scalpel. The tacit knowledge of the surgeon or the cyclist is replicated across all social practices and is reflected in Schön's (1984) account of the reflective practice of professionals. The medical doctors ability to diagnose her patients is not due to 'technical rationality' or the simple application of her knowledge of the medical sciences to the specific patient but a skillful ability built up through extensive experience and training.

The way in which a surgical operation is organised involves a number of individuals (multiple surgeons, anaesthetists, specialist nurses etc) further indicates that, more often than not, practice is a collective, as well as social, activity. The successful operation involves the coordination of a range of tasks performed by number of individuals. This coordination is itself a form or aspect of the individuals practice. We can talk about a surgical team working together like a well-oiled machine and, furthermore, we can consider how replacing a member of the team will change the dynamics of their collective practice. The interrelated nature of practice indicates the degree to which we it would be misguided to place too much focus on the actions of individuals. Praxeological theories of society attempt to mediate between the individualism of, for example, Rational Action Theories and the holism of classical sociology. In Bourdieuan social theory this results in the concept of the social field having a variable focus and is capable of application to micro, median and macro levels.

In micro perspective the social field is constituted by the concrete interactions and practices of individuals. The surgical theatre is a social field constituted by the interactions of the surgeons, anaesthetists, surgical nurses, the patient's body and the various technologies used in the task at hand. The doctor-patient interaction is another such social field. Both of these concrete fields take place within the larger median-level social field of the healthcare center or hospital and, again, these are located within the broader macro-level socio-cultural fields of medicine, nursing and healthcare.

Bottero critiques the Bourdieuan conception and use of social fields as being overly focused "on objective field relations rather than (interpersonal) social relationships" (2009, p. 411). She suggests that there is a sense in which Bourdieu is guilty of "'bracketing off' the concrete nature of social networks as a feature of social space… [and] the variable interactional properties of that space" (Bottero 2009, p. 404). This is, perhaps, the result of the tendency of sociological research to adopt a median level focus. Even when empirically focused on substantive examples of, say, the medical school sociologists are interested in producing more general accounts of the medical school *sui generis*. Thus the concrete social field constituted by specific individuals in specific medical schools becomes abstracted. The conceptual medical school is no longer populated by individuals but by structural positions (professors, non-clinical educators, students, administrators etc.). Nevertheless the conceptual medical school remains, in Bourdieuan terms, a field.

Thompson suggests the 'social field' can be understood as operating at "four semi-autonomous levels: the field of power, the broad field under consideration, the specific field, and social agents in the field as a field themselves" (2008, p. 79). Thus the conceptual 'medical school' of sociological analysis is, in Thompson's terms, a specific field, whilst the actual medical school populated by particular individuals is a 'field as a field in itself'. The medical school also exists within a broader, and multiple, conception of social field, these being the field of medicine and/or healthcare as well as the field of the university or 'higher education.' The field of power is the most abstract of Bourdieuan social fields and it is constituted by the (inter)relation of fields, i.e. how the medical field, the healthcare field, the higher education field and the political field(s) are structured and interact. Whilst the more abstract social fields are often the focus of socio-analysis we must, recall that whatever our focus "if we are to get a proper grasp on the social world [we must] always maintain concrete interactions, where 'it all happens', as an anchor and final point of reference [whilst nevertheless moving] up and down the scale of abstraction" (Crossley and Bottero 2011, p. 116). The perspective articulated here is based on an abstract conception of the UK medical school but, as my focus is on formal educational practices, it keeps in mind that it is specific individuals and their concrete interactions that constitute the medical school and the medical ethics classroom. It is specific individuals who actually teach and learn.

The focus on the actual, and interacting, individual's located within the specific field of the medical school or, even, a medical ethics classroom, is maintained through my primary concern for habitus and its development. Bourdieu's habitus is a generative principle of practice embodied in the individual's dispositions. When combined with a specific social situation (or field) it results in practice. The dispositions of habitus are produced through Bourdieu's 'collective enterprise of inculcation,' the accumulation of experience and an ongoing exposure to the field. As such it is the product of socialisation and, I have suggested, enculturation. Fully formed it entails a practical mastery of the 'game,' 'task' or 'practice' at hand. The metaphor of the game, particularly of sporting games, is one regularly adopted by Bourdieu. Our ability to play games, football or rugby say, is not simply one of knowing and applying the rules but is a matter of our embodied abilities and, furthermore, a practical sense. The football player not only develops skillful abilities with a ball but also a sense of their place in the team and the approach or strategy it takes to playing the game. Relative to their position players have a practical sense of the range of possibilities the field presents. Spectators often develop a similar, but nevertheless different, sense of the possibilities and can, thereby, appreciate (or depreciate) the tactics a team adopts when playing.

This sense of the game and associated ability to perceive the range of possibilities the field presents is produced by the habitus. It should be understood "as a system of dispositions, that is of permanent manners of being, seeing, acting, and *thinking*. Or a system of long-lasting (rather than permanent) schemes or schemata or structures of perception, conception and action" (Bourdieu 2002, pp. 27–28, emphasis added). Some have considered it a deterministic concept (Jenkins 1982) however, consistent with Bourdieu's suggestion that "habitus offers the only form of durable freedom, that given by the mastery of an art, whatever the art" (Bourdieu 1999a, p. 340), Robbins thinks it a form of soft determinism (2012, p. 31) whilst Meisenhelder (2006) likens it to the role of character in earlier social theory. The habitus of medical professionals inculcates within them a set of dispositions which condition their 'manners of being, seeing, acting, and thinking'; it results in their medical gaze and an ability to diagnose patients given the range of possibilities. We can see then that the medical habitus does not determine the diagnosis but, rather, offers the only form of professional freedom, the mastery of the art (and science) of medical practice.

As discussed in more detail in the next section, the medical habitus is produced through the process of medical education, a process that includes aspects of formal education and what we might call an apprenticeship. Through exposure to field of practice the dispositions of individual medical students are altered, developed and restructured to meet the needs of future medical practice. Medical students come to take on the bearing of a Doctor and the ability to see, act, think and practice as one. The development of habitus is an unconscious process but, nevertheless, one which has important conscious effects. In his earlier writings the cognitive aspect of Bourdieu's use of habitus had greater clarity (Lizardo 2004; Maton 2008, p. 57) and one might suggest this made something of a return in his later writing (Bourdieu 1999b, 2000, 2004). Nevertheless most interpreters highlight the non-cognitive aspect of habitus and, therefore, of practice. In the next section I present an overview of some recent Bourdieuan research into medical education before turning to the psychological idea of 'thinking dispositions' to provide a basis to understand habitus as a generative principle of the obviously cognitive practice and pedagogy of professional medical ethics.

2.3 Critical Sociologies of Medical Education

2.3.1 Medical Education Research

Sociological studies of medical education tend to be understood as part of medical sociology. However they also relate to another substantive area of sociological enquiry, the sociology of the professions. For our purposes it is interesting to recall that "if one thing is thought to characterise a profession besides knowledge it is a code of ethics: professionals are people who act ethically and therefore questions of value are the essence of professional practice" (MacDonald 1995, p. 167). However the sociology of the professions often adopts an approach that I earlier termed "irony as methodology" (Anderson and Sharrock 1983). Studies conducted in this mode tend to expose 'moral reversals' which turn the bad into the good or the good into the bad. Such studies tend to undermine or criticise the official discourse to reveal the operation of power, unacknowledged influences and alternative explanations. In this view 'ethics' is often seen as part of the institutional apparatus used by professions to protect their autonomy. Whilst it is certainly the case that the sociological of medical education also has a critical aspect that undermines the discourse of the profession it is nevertheless also the case that it can, and often does, make a positive contribution. Sociological research on medical education offers a more comprehensive understanding of professional reproduction and, particularly, the process of professional socialisation. In so doing the aim is to make a positive contribution to future developments in medical education whether this be in pedagogical terms, in terms of producing better, more 'humanist,' doctors or in regards addressing structural inequalities in terms of class, gender or race.

Whilst there has been relatively little sustained attention paid to the topic since Merton et al.'s Boys in White (1957) and Becker et al.'s Student Physician (1957) there has been recent surge of interest in the area. Whilst some interest was evident in the 1980s (cf: Atkinson 1981 and Colombotos 1988) it was not until the late 1990s the contemporary range of sociological activity in medical education began. Perhaps the pivotal article was Hafferty and Frank's 'The hidden curriculum, ethics teaching, and the structure of medical education' (1994). So influential is the perspective they articulate that the hidden curriculum can now be considered not simply an analytic perspective on medical education but a theory of medical education (Hafferty and Castellani 2009). However the development of sociological perspectives on medical education cannot be credited with responsibility for medical education's contemporary self-understanding. The success of Hafferty and Frank's message must also be considered as a function of changes in the field of medical education, changes that are discussed in the next chapter.

Hafferty and Franks conception of the hidden curriculum seeks to express the fact that assumptions surrounding the informal and formal curriculum, the medical student's accumulation of experience through clinical placements and their classroom-based education, exist in tension. Indeed the informal moral socialisation of medical students that occurs in the clinical context is often presented as a justification for formal ethics education and, latterly, the education in medical humanities. They are understood to offer a humanistic corrective to the techno-scientific excesses of modern medicine that can act to obscure and even dehumanise the individual patient. Whilst Hafferty and Franks highlight the potential for conflict between the morally socialising informal curriculum and the ethical education offered as part of the formal curriculum they offer a deeper analysis concerning the hidden curriculum, defined as "a set of influences that function at the level of organisational structure and culture" (Hafferty 1998, p. 404). The hidden curriculum is presented a major contributor to medical student's socialisation.

As the focus here is primarily on connecting the formal pedagogy of ethics education and the informal moral socialisation of medical students the hidden curriculum is only indirectly addressed. My use of the term enculturation seeks to provide a way to overcome the disconnect between our understanding of the process through which the formal and informal curriculum contributes to the professional reproduction of medical students whilst leaving intact the potential for their content to be in conflict. However, the hidden curriculum should not be considered merely an aspect of medical education's informal curriculum but as also present in medicine's formal curriculum. Once the formal and informal aspects of medical education have been theoretically connected a more constructive and comprehensive analysis of the underlying hidden curriculum should result from further research using this renewed perspective on professional reproduction.

2.3.2 Bourdieuan Studies in Medical Education

Whilst a range of theoretical approaches has been adopted in sociological studies of medical education it is nevertheless the case that Bourdieuan perspectives have predominated. Bourdieuan social theory has been called a "guiding framework" (Brosnan and Turner 2009b, p. 4) for the sociological study of medical education and one that "is primed to examine" Brosnan (2009, p. 60) a range of contemporary questions. Furthermore as a relational sociology it is well placed to overcome the autonomous subsets found within the sociology of medical education that, variously, focus upon and neglect the socialisation of medical students and the structural aspects inherent in the social organisation of medicine and medical education (Hafferty 2000).

Brosnan and Lempp both demonstrate that medical students are required to conform to the structural demands of medical education and, therefore, of the profession. Consistent with the above these demands may be formal, informal or hidden. There are the explicitly recognised requirements of medical education, such as a certain knowledge base, the passing of exams, and a range of informal expectations, conventions and norms. But, as Lempp has demonstrated (2009, p. 79), there is also an requirement to conform to a range of hidden imperatives if the game of medical school is to be played successfully. This indicates the range of inputs that contribute to the formation and reproduction of the medical habitus, something more directly addressed in the work of Sinclair (1997) and Luke (2003).

2.3.3 Developing the Medical Habitus

Brosnan suggests that the "habitus is essentially Bourdieu's theory of socialisation" (2009, p. 56) and, given its intimate connection with the concept of the field and, therefore, social structure it offers a ground for connecting these disparate aspects of the sociology of medical education. In Bourdieuan terms then the socialisation of medical student is a process which develops and (re)structures their habitus through exposure to the broader institutional and organisational culture(s) of medicine. Brosnan's major concern is to articulate the connection between student socialisation and the way in which the profession structures medical knowledge. There is a dichotomy between the art and the science of medical practice and, in her study of 'traditional' and 'innovative' medical education, this (re)emerges as a dichotomy between scientific knowledge and clinical experience with the more humanistic aspects of medical education and knowledge being neglected in both contexts.

In her study of medical education as involving a transition from medical student to pre-registration doctor Lempp (2004, 2009) makes use of Bourdieu's idea of a 'feel for the game'. The idea of social practices as being part of a 'game' is a recurring motif of Bourdieu's work and he often metaphorically compares social life to sport. Differently located individuals struggle for recognition and compete for resources that, in Bourdieu's terms, are understood to be forms of capital (Moore 2008). However, in this context, we can draw a distinction between mastering the practice of medicine, and so 'being a medical professional,' and mastering the game of medical education, and so 'being a medical student.' To varying degrees students arrive at medical school pre-equipped to master the game they are being asked to play. The entry requirements are not simply about grades but also about what kind of person the student is. These two factors are not unrelated, achieving certain grades in certain subjects has some relevance to character. Considering the social context in which this achievement was accomplished can give offer further insights. As Brosnan suggests (2009, p. 61) differing medical schools look for and attract different sorts of applicants. This indicates that the habitus of (future) medical students and professionals is preconditioned by their trajectory through the fields of education and society more generally.[1]

As this indicates we can consider medical professionals, clinical and non-clinical educators, and medical students to embody differing habitus all of which can be implicated in the pedagogical practices of medical education. As such we might consider medical education less a game than a choreographed dance; less a competitive sport, although certainly there are still competitive aspects, and more a cooperative performance. Within the choreography of this performance there remains room for improvisation; pedagogical practices vary between medical schools both in their general typology, i.e. whether they have a Traditional, Integrated and Problem-Based Learning (PBL) curriculum, and in the details of the curriculum and its organisation. Nevertheless there is clearly a repetitive and programmatic aspect to medical education and training. There is a path laid out for

---

[1] It is also interesting to note that whilst there have been significant changes to the demography of medical students in terms of gender and ethnicity over the past four decades there has been little change in terms of their social class (Lempp 2009, p. 73).

students to follow. Some dimensions of this path are explicit and some are not, nevertheless the student must learn to negotiate it if they are to reach the end.

In Bourdieuan terms the 'path' medical students are asked to negotiate is a structured trajectory through the particular social field of medical education. Traversing this path involves the adaptation and transformation of medical student's preexisting habitus (Luke 2003, p. 65) and, as a result, the (re)production and development of a specifically medical habitus. In her study of Australian 'Housemen'—post qualification but pre-full registration junior doctors—Luke prefers to think in terms of professional development rather than socialisation (2003, p. 49). As with my use of the term enculturation her concern is to avoid an 'over-socialised' picture of medical education. However whilst it is legitimate to suggest that, in her study, the medical habitus is fully and empirically present this is not the case in my own. There may still be significant professional development required of Housemen before they can practice independently but, nevertheless, they are engaged in practice in a way that undergraduate medical students are not. In the context of the professionally qualifying degree it is better to restrict oneself to the term socialisation (and enculturation) rather than professional development, as we must consider the acquisition of the professional medical habitus, rather than its further development, the basic teleos of such education. However, we might note that all fall under the rubric of the collective enterprise of inculcation and professional reproduction.

Thus, in a Bourdieuan perspective, the implicit purpose of medical education is the (re)production of the medical habitus. The medical habitus is structurally presented to medical students in formal pedagogy, informal experiences and through role models. Whilst the medical habitus is embodied in many of the medical student's teachers it is important to realise that it is the experience of being taught, the exposure to medical education's various curriculums, and their acclimatisation to the culture(s) of medicine (particularly the medical school and the teaching hospitals) that results in medical student's internalisation of the social and cultural structures of medicine in the form of a specifically medical habitus. Such an understanding reflects Bourdieu's conception of the (medical) habitus as "structured structures predisposed to functions as structuring structures, that is, as principles of the generation and structuring of practices and representations" (Bourdieu 1977, p. 72) and the way in which it is intertwined with the (medical) field, a connection forged through practice (a concept which includes 'rehearsal,' legitimate peripheral participation and apprenticeship i.e. the practice of being a medical student).

In his ethnographic study of medical education 'Making Doctors', Sinclair (1997) attempts to construct an account of the medical habitus or, more accurately, its dispositions. He combines a Bourdieuan approach with the seminal research of Merton et al. who considered socialisation to refer "to the processes through which [the medical student] develops his [sic] professional self with its characteristic values, attitudes, knowledge, and skills, fusing these into a more or less consistent set of dispositions which govern his behaviour in a wide variety of professional (and extra-professional) situations" (Merton 1957, p. 287). As the habitus is considered to be "systems of durable transposable dispositions" (Bourdieu 1992, p. 53) there is a prima facie compatibility with this earlier 'dispositional' perspective.

Sinclair identifies eight dispositions: Competition; Co-operation; Economy; Experience; Idealism; Knowledge; Responsibility; and Status. Sinclair consider Knowledge, Experience and Responsibility to be the primary dispositions of the medical habitus and, given the importance of all three to professional medical practice based on specialisms largely determined by the biomedical sciences, we can see why this might be. Whilst Sinclair's dispositions are a useful characterisation of the medical habitus they only provide a general indication of the practice(s) they are supposed to underpin. As my project here is more tightly focused on medical ethics education and, therefore, on the practice of medical ethics greater specificity regarding the dispositions being enculturated is required. Given the nature of contemporary medical ethics as form of applied philosophy—a cognitive practice—this specification requires, first, a recognition of the cognitive aspects of habitus and, second, the appropriation of the psychological concept of 'thinking dispositions.'

2.4 The Cognitive (Medical) Habitus Needs (Ethical) Thinking Dispositions

That Bourdieu offers a way to properly recognise the unconscious aspects of our social practices is central to the power of his social theory. However we must also recognise that many of our practices have a cognitive dimension. At minimum Bourdieu tends to underemphasised the cognitive aspects of practice. Nevertheless within a Bourdieuan perspective there are strong reasons to acknowledge that communication and reflection are a necessary aspect of acquiring even the most embodied forms of practice (Nobel and Watkins 2003). To suggest that certain of our practices are cognitive is not to suggest that they are asocial or that they occur in a decontextualised or acultural manner. The practice of ethics, or of ethical reflection, is an aspect of modern medical practice and so it is an example of the social and cultural nature of our cognitive practices. Whatever one thinks of the philosophical foundations of medical ethics, or of morality and ethics more generally, one must recognise that, if it is to be practiced, the culture of medicine is central to the nature of contemporary medical ethics, and vice versa. Any philosophical medical ethics must be instantiated within medical culture. As such professionals medical ethics is a 'way of thinking' consistent with 'how doctors think' (Montgomery 2005) more generally. We might also point to the relevance of the way medicine constructs its objects (Good 1993, Chap. 3) and, particularly, the language of case presentation (Anspach 1988) for the discourse of medical ethics.

The primary characteristic of contemporary medical ethics is, perhaps, its cognitive or reflective nature. Prior to the emergence of bioethics the ethical practice of medicine was considered to be largely intuitive and a function of the kind of (gentle)men, and it was broadly men, that doctors were. It is perhaps best summed up by an undated quotation that prefaces the IME's Pond Report. Attributed to 'A Doctor' it reads: "One is ethical—I mean, for heaven's sake—is one not?" (Pond 1987, Preface citation). Until relatively recently it was assumed that medical professionals would simply know what the right thing to do was and, furthermore, could be unquestionable relied upon to do it. As discussed in the following chapters the emergence of reflection in medical ethics was accompanied, if not preceded, by the emergence of reflection across other aspects of the medical curriculum and, indeed, by the very idea of the professional as a reflective practitioner (Schön 1984). The 'reflective' nature of modern medical ethics is not simply a function of external applied (bio)ethical discourses but echoes other developments in the culture of medicine and medical education. As such medical ethics can be considered as one aspect of the broader medical habitus of the reflective professional.

It can seem odd to consider 'reflection' as an aspect of habitus. As Brubaker points out such a perspective implies that there are "unreflective dispositions to reflect" (1993, p. 225) suggesting that, in the relevant contexts, those with such dispositions automatically engage in reflection. Furthermore we reflect and act on our beliefs, something that modern theorists such as Ryle have characterised as dispositional (Price 1969, p. 243 see also Schwitzgebel 2002). Stueber (2005) has sought to explore the notion of dispositions in the context of following rules and, therefore, beliefs. His analysis differentiates between first and second order dispositions. The former involve the production of regular of behaviour and can include reflective activity whilst the latter produce reflective activity that monitors and can affect current and future behaviour. Schön proposes that reflective practice is an essential aspect of what professionals do whilst Bourdieu considers reflexivity to be a (normative) aspect of sociological habitus. Thus we should consider reflection and reflexivity to be supported by the relevant dispositions and, in the latter case, predicated on the sociologist's social scientific habitus (Bourdieu 2004, p. 89). The conclusion that the cognitive activities of researchers and professionals are underpinned by the thinking dispositions of the cognitive habitus is unavoidable.

The idea of 'thinking dispositions' is prevalent in the more psychological strands of education and philosophy but has not made any real impact in sociology, including cognitive sociology. One might question whether it is valid to adopt this predominantly 'psychological' concept under a broadly sociological rubric. The answer turns on what, precisely, is meant by a disposition, something Bourdieu did not discuss extensively. He presents his use of the term disposition as being in the interests of provoking "a more concrete intuition of what habitus is and to remind you what is at stake in the use of such a concept, namely a peculiar philosophy of action, or better, of practice, sometimes characterised as *dispositional*" (Bourdieu 2002, pp. 27–28). In a widely cited footnote he suggests that:

> "The word disposition seems particularly suited to express what is covered by the concept of habitus (defined as a system of dispositions). It expresses first the result of an organising action, with meaning close to that of words such as structure; it also designates a way of

being, a habitual state (especially that of the body) and, in particular, a predisposition, tendency, propensity, or inclination. [The semantic cluster of 'disposition' is rather wider in French than in English, but as this note—translated literally—shows, the equivalence is adequate. Translator.]" (Bourdieu 1977, p. 214, fn1).

Interestingly this exact phrasing—a tendency, propensity, or inclination—has been used to characterise thinking dispositions; they are "a tendency, propensity, or inclination, to think in certain ways under certain circumstances" (Siegel 1999, p. 209). Whilst there is a prima facie affinity between Bourdieu's social theory and this idea of thinking dispositions we might still be concerned that this offers a deterministic picture of human agency, a charge often leveled at Bourdieu and 'dispositionalism' (Lahire 2010, pp. 50–56). However consistent with Swartz's suggestion that Bourdieu's "[a]ctors are not rule followers or norm obeyers but strategic improvisers who respond dispositionally to the opportunities and constraints offered by various situations" (Schwartz 1997, p. 100) Siegel suggests that thinking dispositions cannot legitimately be thought of as deterministic, concluding "that thinking dispositions are not reducible to either formal rules of thought or to particular behaviours or patterns of behaviour" (Siegel 1999, p. 214).

The conceptualisation of thinking dispositions has an affinity with the idea of enculturation and with the acquisition/participation metaphors of learning. Tishman and Andrade suggest they "are learned through a process of enculturation rather than direct transmission. Thinking dispositions… are characterological in nature, and, like many human character traits, they develop in response to immersion in a particular cultural milieu" (1996, p. 9). Thinking dispositions are not just one thing, there is no singular way of thinking well. Instead the concept of thinking dispositions reveals that we should consider there to be better and worse 'ways of thinking' in particular situations or fields. Thus we might, again, distinguish between not only better and worse ethical thinking but better and worse ethical thinking in the cultural milieu of professional medical practice as compared to the field of applied philosophy or 'academic bioethics'. Thus whilst Perkins et al. (1993) identify seven ideal thinking dispositions there is a need to give further specification a pros pos the cultural milieu of medical ethics education and, implicitly, medical ethical practice.[2] In what follows I take each in turn and consider its relevance for professional medical ethical thinking and the medical ethics education that aims to foster them.

---

[2] It is interesting to note that Ennis (1996) offers a set of 14 thinking dispositions that are more detailed and less generalised than those discussed here. Furthermore Tishman and Andrade (1996), p. 5 compare thinking dispositions to Costa's (1991) five 'passions of mind,' these being Efficacy, Flexibility, Craftsmanship, Consciousness, and Inter-dependence. In my view these are reminiscent of Sinclair's (1997) use the concept of dispositions.

2.4.1 The Disposition to Be Broad and Adventurous

A thinking disposition to be broad and adventurous is unlikely to be a central medical or ethical thinking disposition. In a reasonably strong sense medical education, socialisation and enculturation is aimed at reproduction, i.e. the production of conformity. Whilst medical students testing medical knowledge on themselves (Sinclair 1997) is indicative of a certain sort of adventurousness and a broad or 'generalist' thinking disposition may well be required of those who intend to go into General Practice. Nevertheless it is the case that medical professionals should be morally and epistemically interchangeable. In the case of both medical diagnosis and ethical judgement it is conformity, not adventurousness, that is of primary value in practice.

2.4.2 The Disposition Toward Sustained Intellectual Curiosity

Undergraduate medical education is only the first step in a career that now involves continuing professional development. In this sense the medical habitus must include a disposition towards sustained intellectual curiosity. However, this disposition may only variably apply to medical ethics. Medical practice demands a practical disposition and a certain level of ethical intellectualism is not consistent with the imperative to act. Whilst there is increasing opportunities for medical students and professionals to pursue their intellectual curiosity as with other areas of medicine there is an ethical division of labour. Those whose curiosity extends to ethics can pursue this interest and, in doing so, make a distinctive contribution to the profession. In particular we might highlight the contribution they might make to the institutionalised ethical discourse of this self-governing profession. However there is a point at which developing a disposition towards sustained intellectual curiosity obstructs the production of practicing professionals. Thus its development in the context of undergraduate medical education might be delimited or highly directed.

2.4.3 The Disposition to Clarify and Seek Understanding

Seeking clarity and understanding is central to the medical habitus. A doctor must, for example, clarify symptoms with patients and make sure they properly understand what is being communicated. An applied philosophical view of medical ethics indicates that this disposition is likely to produce a concern for the definition of ethical concepts and their appropriate application to (generalised) cases. This disposition motivates discussion and reflection on the ethical issues of medicine and the identification, resolution or accommodation of disagreement. However one might be concerned that, within the medical habitus, clarity and understanding is sought in such a way that patient's individuality can be obscured. An overly abstract view of medical ethics is common. Nevertheless, suitable orientated, this disposition can provide a link to the affective and emotional aspects of medical ethics and the doctor-patient relationship.

2.4.4 The Disposition to Be Planful and Strategic

Certainly the medical habitus must include a disposition to strategically plan the care of patient(s), particularly in the context of multi-professional teams. In the case of medical ethics this is perhaps best represented by prospective discussions of Do Not Attempt Resuscitation (DNAR) orders. This can be considered in relation to specific patients but also in terms of wider biomedical discourses. There has been a recent trend towards questioning the presentation of resuscitation in medical dramas in order to counter the public's assumptions regarding the success rates of the procedure. Encouraging the patients to consider their wishes in a range of circumstances, particularly with regard to organ donation, is another aspect of the strategic planning of medical ethics. The prospective discussion of ethical principles during medical education, and the formation of professional ethical guidance by the GMC, the BMA etc. is also an example of this disposition.

2.4.5 The Disposition to Be Intellectually Careful

One can see the enculturation of a disposition to be intellectually careful in the idea of medical education as involving training for uncertainty (Fox 1957) and certainty (Atkinson 1984). The uncertainty of medical practice is often hidden from patients who can often be presented with certainties that are, nevertheless, inherently probabilistic and concern the management of risks. Care is taken not to unnecessarily concern patients whilst also not misleading them. The range of possible diagnosis of a set of symptoms or a proposed test might only be alluded to. Modern medical ethics is often construed as theoretically requiring full disclosure however a more nuanced, or intellectually careful, understanding reveals it requires a sensitivity to the patient and the circumstances, including the medical and the social circumstances. We might note that this disposition should, as with the medical habitus more generally, be understood as being fundamentally conditioned by the practical, rather than theoretical, dimensions of both scientific and ethical practice.

2.4.6 The Disposition to Seek and Evaluate Reasons

The imperative to seek and evaluate reasons is certainly an aspect of a medical ethical thinking disposition. However, in a strong sense medical ethics education is not about teaching medical students solve ethical dilemmas as much as it is about teaching them to have reasons for their actions. This is a subtle and, over time, somewhat blurred distinction. Medicine, and particularly certain of the medical specialisms, have characteristic ethical problems and associated 'solutions' or responses. Thus the ethics of medicine involves stock reasons that are 'pre-evaluated' both in the sense that they are institutionalised in codes or guidance or through repetition in practice. In the case of ethics this disposition may not always be fully exercised in response to characteristic cases but may come to the fore when cases deviate in ethically important ways. Furthermore it is obviously relevant to the ongoing ethical discourses of the profession and its interlocutors where the characteristic case, its deviations and the reasoned responses of medical professionals are more fully discussed.

2.4.7 The Disposition to Be Metacognitive

It is certainly the case that a metacognitive disposition or, perhaps, what Schön (1984) would call reflection-on-action is central to medical education and practice. However, as discussed in Chap. 1, what metacognition denotes varies across differing disciplines, notably (analytic) philosophy and theories of (professional) practice. For the purposes of this book metacognition can be considered reflection that involves relating abstract concepts and concrete circumstances in a dialogical and goal related context. Thus, in practice, this metacognitive disposition involves ethically discussing cases in terms of the abstract concepts and principles of medical ethics. In characteristic cases this may be a relatively simple task but in more unusual or complex cases this may involve more extensive reflection and analysis. It is also central to the broader professional discourse on ethics and its codification and official guidance.

2.5 Conclusion

In this chapter I have offered a brief introduction to Bourdieuan social theory and an overview of some recent research into medical education that has been conducted under its rubric. I have argued that if we are to adequately understand the ethical enculturation of medical students, as well as their moral socialisation, then we must develop a more cognitive perspective on habitus ideally with reference to the idea of thinking dispositions. It is evident that the professional practice of medical ethics is a cognitive and reflective activity but that it differs from the academic practice of applied ethics. As such we should recognise that primarily normative aspect of bioethics does not lie in the any 'solutions' to ethical questions it may offer but in regards the way ethics ought to be done. In the course of medical practice these prescriptions can only be partially fulfilled and the insights of applied ethics as an academic discipline must be culturally reappropriated into the professional medicine context(s). Although it is perhaps not always fully recognised academic bioethicists can and do make a contribution to this cultural reappropriation.

This reappropriation results in a renewed, similarly normative, framework for 'doing' medical ethics. I have presented this framework in terms of thinking dispositions and it should be recognised that these are an idealisation of what happens in medical practice. As such, in the context of research such as my own, the construction of the medical habitus is also an idealisation; it reflects not only how medicine is practice but works with such perspective to articulate how medicine, or in this case medical ethics, should be practiced. My articulation of the medical ethical thinking dispositions builds on the understanding of medical practice offered by Bourdieuan social theory and medical education research that makes use of it. It offers an account of how medical ethics can be practiced given what we know about how professionals in fact practice. If we accept that ought implies can then we can see how the second order normative perspectives of applied (bio)ethics must, and indeed are, be modulated to meet the exigencies of medical practice.

The perspective I have offered recognises that the purpose of medical ethics education is not merely to impart the ability to think or act in certain ways but also seeks to develop the disposition to do so in certain circumstances. This view "goes beyond a skills-centered view and proposes a dispositional approach to the teaching of thinking.... [which] means more than inculcating particular thinking skills; it means teaching students to be disposed to think creatively and critically in appropriate contexts" (Tishman et al. 1993, p. 147). At its best the enculturation of medical ethics should compliment the moral socialisation of medical students and result in professionals who intuitively behave in an ethical manner and, furthermore, as part of this intuitive behaviour engage in creative and critical ethical reflection when and where the circumstances require them to do so.

References

Albert, M., and S. Reeves. 2010. Setting some new standards in medical education research. *Medical Education* 44(7): 638–639.

Anderson, D., and W. Sharrock. 1983. Irony as a methodological theory: A sketch of four sociological variations. *Poetics Today* 4(3): 565–579.

Anspach, R.R. 1988. Notes on the sociology of medical discourse: The language of case presentation. *Journal of Health and Social Behavior* 29(4): 357–375.

Atkinson, P. 1981. *The clinical experience: The construction and reconstruction of medical reality.* Guildford: Gower Publishing Ltd.

Atkinson, P. 1984. Training for certainty. *Social Science and Medicine* 19(9): 949–956.

Becker, H.S., B. Geer, E.C. Hughes, and A.L. Strauss. 2007. (First Published: 1961). *Boys in white: Student culture in medical school.* London: Transaction Publishers.

Bottero, W. 2009. Relationality and social interaction. *The British Journal of Sociology* 60(2): 399–420.

Bottero, W., and N. Crossley. 2011. Worlds, fields and networks: Becker, Bourdieu and the structures of social relations. *Cultural Sociology* 5(1): 99–119.

Bourdieu, P. 1977. *Outline of a theory of practice.* Cambridge: Cambridge University Press.

Bourdieu, P. 1992. *The logic of practice.* Cambridge: Polity Press.

Bourdieu, P. 1999a. Scattered remarks. *European Journal of Social Theory* 2(3): 334–340.

Bourdieu, P. 1999b. *The weight of the world: Social suffering in contemporary society.* USA: Polity Press.

Bourdieu, P. 2000. *Pascalian meditations.* Cambridge: Polity Press.

Bourdieu, P. 2002. Habitus. In *Habitus: A sense of place*, eds. Hillier, J. and E. Rooksby, 27–36. Aldershot: Ashgate.

Bourdieu, P. 2004. *Science of science and reflexivity.* Cambridge: Polity Press.

Bourdieu, P., and L. Wacquant. 1992. *An invitation to reflexive sociology.* Great Britain: Polity Press.

Brosnan, C. 2008. The sociology of medical education: The struggle for legitimate knowledge in two English medical schools (unpublished Ph.D. thesis). University of Cambridge.

Brosnan, C. 2009. Pierre Bourdieu and the theory of medical education: Thinking 'Relationally' about medical students and medical curricula. In *Handbook of the sociology of medical education*, Brosnan, C and B.S. Turner, 51–68. NY: Routledge.

Brosnan, C. and B.S. Turner. 2009b. Introduction: The struggle over medical knowledge. In *Handbook of the sociology of medical education*, Brosnan, C & B.S. Turner, 1–12. Abingdon: Routledge.

Brosnan, C. 2010. Making sense of differences between medical schools through Bourdieu's concept of 'field'. *Medical Education* 44(7): 645–652.

Brosnan, C., and B. Turner. 2009a. *Handbook of the sociology of medical education.* London: Routledge.

Brubaker, R. 1993. Social theory as habitus. In *Bourdieu: Critical perspectives*, eds. Calhoun, C., E. LiPuma, and M. Pstone, 212–234. Cambridge: Polity Press.

Colombotos, J. 1988. Introduction to the special issue: Continuities in the sociology of medical education. *Journal of Health and Social Behavior* 29(4): 271–278.

Costa, A.L. (ed.). 1991. *Developing minds: Programs for teaching thinking.* Virginia: Association for Supervision and Curriculum Development.

Ennis, R.H. 1996. *Critical thinking.* Englewood Cliffs: Prentice Hall.

Fox, R.C. 1957. Training for uncertainty. In *The student physician*, eds. Merton, R. G.G. Reader, and P.L. Kendall, pp. 207–241. Cambridge: Harvard University Press.

Good, B.J. 1993. *Medicine, rationality and experience: An anthropological perspective.* Cambridge: Cambridge University Press.

Grenfell, M. (ed.). 2008. *Pierre Bourdieu: Key concepts.* Durham: Acumen.

Hafferty, F.W. 1998. Beyond curriculum reform: Confronting medicine's hidden curriculum. *Academic Medicine* 73(4): 403–407.

Hafferty, F.W. 2000. Reconfiguring the sociology of medical education: Emerging topics and pressing issues. In *Handbook of medical sociology*, eds. Bird, C.E., P. Conrad, and A.M. Fremont, 238–257. Upper Saddle River: Prentice Hall.

Hafferty, F.W. and B. Castellani. 2009. The hidden curriculum: A theory of medical education. In *Handbook of the sociology of medical education*, eds. Brosnan, C and B.S. Turner, 51–68. Abingdon: Routledge.

Hafferty, F.W., and R. Franks. 1994. The hidden curriculum, ethics teaching, and the structure of medical education. *Academic Medicine* 69(11): 861–871.

Jefferys, M., and M.A. Elston. 1989. The medical school as a social organization. *Medical Education* 23(3): 242–251.

Jenkins, R. 1982. Pierre Bourdieu and the reproduction of determinism. *Sociology* 16(2): 270–281.

Lahire, B. 2010. The plural actor. Cambridge: Polity Press.

Lempp, H.K. 2004. Undergraduate Medical Education: A Transition from Medical Student to Pre-Registration Doctor (unpublished Ph.D. thesis). Goldsmiths College, University of London.

Lempp, H. 2009. Medical-school culture. In *Handbook of the sociology of medical education*, eds. Brosnan, C and B.S. Turner, 71–88. London: Routledge.

Lizardo, O. 2004. The cognitive origins of Bourdieu's habitus. *Journal for the Theory of Social Behaviour* 34(4): 375–401.

Luke, H. 2003. *Medical education and sociology of medical habitus: It's not about the stethoscope!*. The Netherlands: Kluwer Academic Publishers.

Macdonald, K.M. 1995. *The sociology of the professions*. London: Sage.

Maton, K. 2008. Habitus. In *Pierre Bourdieu: Key concepts*, ed. M. Grenfel, 49–65. Stocksfield: Acumen.

Meisenhelder, T. 2006. From character to habitus in sociology. *The Social Science Journal* 43(1): 55–66.

Merton, R.K., P.L. Kendall, and G.G. Reader (eds.). 1957. *The student-physician*. Cambridge: Harvard University Press.

Montgomery, K. 2005. *How doctors think: Clinical judgement and the practice of medicine*. New York: Oxford University Press.

Moore, R. 2008. Capital. In *Pierre Bourdieu: Key concepts,* ed. Grenfell, M, 101–117. Stocksfield: Acumen Publishing.

Noble, G., and M. Watkins. 2003. So, how did Bourdieu learn to play tennis? Habitus, consciousness and habituation. *Cultural Studies* 17(3): 520–539.

Norman, G. 2011. Fifty years of medical education research: Waves of migration. *Medical Education* 45(8): 785–791.

Perkins, D.N., E. Jay, and S. Tishman. 1993. Beyond abilities: A dispositional theory of thinking. *Merrill-Palmer Quarterly: Journal of Developmental Psychology* 39(1): 1–21.

Pond, D. 1987. *IME report of a working party on the teaching of medical ethics (the pond report)*. London: The Institute of Medical Ethics.

Price, H.H. 1969. *Belief: The Gifford lectures*. London: George Allen & Unwin Ltd.

Robbins, D. 1993. The practical importance of Bourdieu's analyses of higher education. *Studies in Higher Education* 18(2): 151–163.

Robbins, D. 2012. La Philosophie et les Sciences Sociales: Bourdieu, Merleau-Ponty et Husserl." *Cités.* 51(3):1731. (In English translation available from the author).

Schön, D.A. 1984. *The reflective practitioner: How professionals think in action*. USA: Basic Books.

Schwartz, D. 1997. *Culture and power: The sociology of Pierre Bourdieu*. Chicago: University of Chicago Press.

Schwitzgebel, E. 2002. A phenomenal, dispositional account of belief. *Noûs* 36(2):249–275.

Siegel, H. 1999. What (good) are thinking dispositions? *Educational Theory* 49(2): 207–221.

Sinclair, S. 1997. *Making doctors: An institutional apprenticeship*. Oxford: Berg Publishers.

Stueber, K.R. 2005. How to think about rules and rule following. *Philosophy of the Social Sciences* 35(3): 307–323.

Thompson, P. 2008. Field. In *Pierre Bourdieu: Key concepts*, ed. Grenfel, M, 67–84. Stocksfield: Acumen.

Tishman, S. and A. Andrade. 1996. *Thinking dispositions: A review of current theories, practices, and issues*. Cambridge: Project Zero, Harvard University. Available from: http://learnweb.harvard.edu/alps/thinking/docs/dispositions.htm.

Tishman, S., E. Jay, and D.N. Perkins. 1993. Teaching thinking dispositions: From transmission to enculturation. *Theory into Practice* 32(3): 147–153.

Chapter 3
Medical Ethics as an Aspect of Medical Education: A UK Perspective

3.1 Introduction: The Rise of Modern Medical Ethics in the UK

The development of medical ethics in the UK has been closely tied to its development as an aspect of medical education. Whilst the development of bioethics in America was marked by the activities of medical outsiders and a concern with broader bioethical questions in the UK it has been marked by 'insiders' or, more accurately, by 'insider–outsiders' and by a predominant concern for the *professional* issues of medical ethics. Some of this can be seen in the work of Maurice Pappworth, an early medical ethical whistleblower often considered the UK's Henry Beecher. As well as raising concerns about unethical practices in biomedical research Pappworth also raised concerns about certain pedagogic practices, notably the internal examination of sedated women without obtaining their consent. However the link between medical ethics and medical education is most clear in the advent of the London Medical Group (LMG), the export of this model to other medical schools in Great Britain and its role in the creation of the Institution of Medical Ethics (IME).

The history of the LMG has been reconstructed by Whong-Barr (2003; see also: Shotter 1988), however it is worth recapping the main points here. The LMG was created by Rev. (later the Very Rev.) Edward Shotter. As a Chaplin to the University of London he had pastoral responsibility for medical students across the city's teaching hospitals. He was also intercollegiate secretary to the Student Christian Movement (SCM) in London (Reynolds and Tansay 2007, p. 189) and commenced his duties around the same time as a report, commissioned by the SCM, on the implications of studying medicine for students and their pastoral care was published (Mepham 1963).

The report suggested that, particularly in the latter years of their education, medical students should be seen as more part of the teaching hospitals than the university and, further, that care should be taken to ensure that patient was not lost

N. Emmerich, *Medical Ethics Education: An Interdisciplinary and Social Theoretical Perspective*, SpringerBriefs in Ethics, DOI: 10.1007/978-3-319-00485-3_3, © The Author(s) 2013

amid wonders of modern medicine.[1] It "indicated [that there was] a need for a service dealing with issues raised through the practice of medicine which concern the theologian, philosopher and sociologist, as well as the doctor" (Reynolds and Tansey 2007, p. 7, fn15). Subsequently Shotter was asked to look into medical education by Bishop Reeves (Reynolds and Tansey 2007, p. 77) where upon he was surprised to find that there was no teaching of medical ethics and that "despite the popular mythology of doctors taking the Hippocratic Oath... there was no such thing" (Reynolds and Tansey 2007: 7, p. 77). Subsequently, with input from 'two or three' medical students, he began to hold lectures on medical ethics under the auspices of the SCM. One of the first lectures, on the nature and management of terminal pain, was given by Dr Cicely Saunders (later Dame Saunders), who had been an advisor to Mephem (Reynolds and Tansey 2007, p. 7) and is credited with founding the hospice movement. In the first year the number of lectures was four, in the second year eight and in the third year twenty-one. By 1966 it was felt that the LMG, as it had became to be known, had outgrown the SCM and it became an independent body.

In time other medical schools and their associated teaching hospitals took up the LMG model and whilst most were initiated and run by university chaplains it is clear that these discussions groups were not only an ecumenical but also professional and secular exercise.[2] Those who expressed concern about the religious constitution of the LMG were often co-opted onto its board and efforts were made to ensure it was not seen as a 'chaplaincy exercise' by the profession (Reynolds and Tansey 2007, p. 10). Indeed, announcing himself as a Jewish-Atheist, Raanan Gillon recalls being welcomed, albeit as a *Godsend*, by the Rev Shotter (Gillon 2005, p. 90). Overtime the LMG's activities generated a 'postgraduate advisory group' that subsequently became the Society for the Study of Medical Ethics (SSME) and, in turn, became the IME in 1984.

The early development of the IME in the UK cannot be directly compared to that of the Hastings Centre and the Kennedy Institute of Ethics (KIE) in the USA. In the first instance whilst it was foreshadowed by the LMG and the SSME the IME was not officially formed until the late 1980s whereas the American centers were founded in 1971 and 1969 respectively. Furthermore the IME is much less independent of the medical profession than the Hastings Centre and it is not part of an academic institution like the KIE. At its inception the LMG, and similar groups, were unofficial and 'extra-curricula' but were, nevertheless, attached to medical schools and teaching hospitals. Thus the early development of medical ethics in the UK can be contrasted with the development of Bioethics in America where predominantly theologically inclined outsiders inserted themselves into medical

---

[1] The relevant passages of the Mepham Report were cited in the Annual report of the LMG 1965 and 1967/8. See: Reynolds and Tansey (2007, pp. 7–8 and 76–77).

[2] Shotter indicates that the primary reason the other medical groups were convened by clerical, rather than medical, professionals was the pragmatic fact that his 'bush telegraph' was comprised of fellow ministers involved in pastoral care of (medical) students (Reynolds and Tansey 2007, p. 10).

schools and hospitals and then managed to receive 'official' appointments (Jonsen 1998). These individuals soon became medical insiders and their *bioethics* a part of the official curriculum. In contrast the LMG developed within the context of medical education, clinical practice and pastoral concern but, for a significant period of time, its *medical ethics* remained an extra-curricula pursuit. Another indication of its insider–outsider status. It is clear that whilst the Rev. Shotter provided an organisational focus for the LMGs activities they were conducted by and for medical professionals and students.[3] During its long gestation UK medical ethics was a niche interest, more peripheral than its American counterpart and a place where those involved can often be considered 'insider–outsiders' of some kind.

Recently there seems to have been a desire to tell the history of medical ethics in the UK motivated, in part, "to provide a corrective to accounts from across the pond" (Reynolds and Tansey 2007, p. 4). This thought echoes Dunstan's address to an Anglo-American conference on bioethics held in New York in 1986. Here Dunstan drew attention to the way in which the Church of England had sought to address medical ethical issues, including 'Dying Well' (Boyd 2009, p. 488). Furthermore it had had a positive involvement in the development of the 1967 Abortion Act and in the way in which a tradition of internal ethical debate had developed within the UK medical profession. In contrast the bioethical debate (something slightly different to medical ethical debate) in America had been led by medical outsiders and, Dunstan suggested, is more 'showy' (1988, p. 5). Nevertheless, the 1980s had certainly seen an increase in bio- and medical ethical activity in the UK. The GMC held a conference on the teaching of medical ethics in 1984 and the papers were published by the JME the following year (Walton 1985 and papers in same issue). A similar collection was also published in 1987 (Gillon 1987 and papers in issue). Having presented a number of BBC programs on medical ethical issues Ian (later Sir Ian) Kennedy gave the Reith Lectures in 1980 (Wilson 2011a). These were published under the title 'Unmasking Medicine' in 1981 and, having occasioned widespread comment and debate both within and without the profession, were revised and reissued in 1983. 1982 saw the formation of the Warnock Committee tasked with inquiring into the ethical aspect of human fertilisation and embryological science and technologies (Wilson 2011b).[4] It is clear that this Anglo-American Dialogue between, essentially, representatives from the Royal Society of Medicine (RSM) and the Hasting's Centre it is the case that British medical ethics had come to learn from American Bioethics.

---

[3] Shotter suggests that the discussions of the LMG were open to the public and whilst this may have been the case the fact was that they were advertised on the notice boards of the London hospitals meant that, in effect, discussions remained in camera.

[4] It is interesting to note that Wilson suggests this period of the 1980s should not be mythologically construed as the historical origin of UK bio- and medical ethics but, rather, its high-water mark, presumably in terms of impact on government policy and the structural aspects of the medical profession and, we might add, medical ethics education (2011a).

A few years prior to this conference, in 1984, the then editor of the JME, Dr Ranaan Gillon,[5] had, with sponsorship from the Medicine Gilliland Foundation via the Royal College of Physicians (RCP), travelled to America with a not dissimilar purpose (Gillon 2005, p. 92). Gillon is a GP who developed an interest in medical ethics during his undergraduate degree and his purpose was to learn about the teaching of medical ethics in American Medical Schools (see Gillon 1990). Having qualified in 1964 he had found himself unable to formally pursue his interest, being told by Sir George Pickering, Reguis Professor of Medicine at Oxford, that medical ethics was not a subject that one could study as "one acquired it in the course of learning to become a good doctor [and i]f by chance medical ethics were amenable to study it could only usefully be studied by mature doctors with a wealth of clinical experience upon which they could draw" (Gillon 2005, p. 86). Having spent time as a journalist for the Medical Tribune Gillon did eventually pursue his interest via an undergraduate degree in philosophy at Birkbeck College (University of London), later giving up his doctoral studies at an early stage to take up the editorship of the JME.

Given his professional trajectory we can consider Gillon another example of the insider-outsider found within UK medical ethics. He has been at the forefront of developments in UK medical ethics and, particularly, medical ethics education since the late 1970s when, acting as lynchpin between the Oxford University philosopher Dr Michael Lockwood and the audience which would have been largely comprised of philosophically untrained medical professionals, he taught on the UK's first formal course in medical ethics. His book 'Philosophical Medical Ethics' (Gillon 1986), based on a series of articles published in the BMJ in 1985 and 1986, continues to be an influential text within the UK profession, although the long promised revised edition remains in the pipeline. Whilst the 'four principles approach' is somewhat disdained by academic bioethics in the UK it remains the pre-eminent approach within professional practice and education, where Gillon remains its leading advocate and interpreter (Boyd 2009, p. 488). He has been a central figure in the IME and involved with the development of the core curriculum in medical ethics first published in 1998 (Anon, JME) and revised in 2010 (Stirrat et al.). Whilst the first iteration of the consensus statement immediately references Tomorrow's Doctors (Anon, JME, p. 188, fn.1) and the second iteration explicitly claims in its title that it is 'for doctors of tomorrow' (Stirrat et al. 2010, p. 55) the precursor of these statements in fact predates 1993s Tomorrow's Doctors.[6] The first 'consensus statement' regarding the medical ethics curricula

---

[5] Dr Gillon took over from the founding editor, Dr Alastair V. Campbell who published the first UK book on medical ethics in 1972. Campbell and Gillon were jointly awarded the Henry Knowles Beecher prize for contributions to ethics and the life sciences by the Hastings Centre. See: http://www.thehastingscenter.org/About/Default.aspx?id=2972 [Accessed June 2011].

[6] Tomorrow's Doctors is a document first published by the GMC in 1993 and revised in 2003 and 2009. It sets the agenda for undergraduate medical education in the UK and has proved to be an extremely influential document. It is discussed further below.

issued by the IME is the Pond Report (1987) that, with financial support from the Nuffield Foundation, it commissioned and published.

The Pond Report is a short but reasonably comprehensive document that considers the nature of medical ethics, whether or not it can be taught, its scope and applicability to a variety of medical arenas including the various medical specialties, nursing, research and management. It contains a survey of medical Deans and medical students regarding existing arrangements for ethics education and concludes with some broad recommendations for future activities. An appendix considers the specific contributions to teaching that theology, philosophy and law can make. The report refrains from advocating for certain kinds of pedagogic activities or teaching arrangements or, as later promoted by the IME, from specifying a 'core curriculum.' The only advice advanced with certainty is the inadvisability of allowing teachers who are ethically 'certain' to teach medical ethics. Which is to say that the advice for teachers of medical ethics to medical students is to adopt a pluralistic approach, being certain to avoid taking a hortatory, or 'moralising', stance in regards particular issues.

This advice is particularly interesting. As well as being a secular enterprise medical students are not to be imparted with determinate moral solutions, answers or positions by those who are morally certain and, therefore, cannot but adopt a 'moralising' or 'hortatory' pedagogic style. Rather the aim is to facilitate the (medical) ethical development of medical students and encourage them to explore the moral aspects of medical practice. This reflects a new, and still contemporary, ideal of the secular professional: an individual doctor who is morally unbiased and objective. In the context of the secular public sphere there is no room for previous professional ideals based on the English, Church of England (Non-Catholic, Non-Jewish),[7] Gentleman (Pond 1987, p. 38). Whilst this secular ideal is, of course, as much of a moral ideal as those held by previous generations its normative dimension is not to be found in determinate ethical opinions or positions but in an openness to moral discourse and dialogue. However, this discourse is normatively constrained, as it must be of a particular style. Rather than relying on the judgements of a particular type of persons the normativity of modern medical ethics is 'meta-ethical.' Modern medical ethics is marked by a specific way of thinking: a process of abstraction, which is broadly reliant on principles, cases and casuistic reasoning.

In this regard we can agree with Gillon's suggestion that UK or 'British' medical ethics is typically characterised by "a pragmatic, situationist, common-sense, anti-theoretical, and anti-regulatory approach [... and whilst] it is gradually being acknowledged that some theoretical underpinning is needed even for common-sense ethical decisions ... virtually all involved in the British medical ethics scene agree on one issue: the central importance of real cases, manifesting real

---

[7] The early professional biography of Maurice Pappworth stands as a testament to the anti-Semitism of British society of the time (Booth 1994). The anti-Catholic nature of a certain strata of English society is well established.

medico-moral problems, in their real human context, for an adequate critical study, teaching, or understanding of the 'humanized version of ethics'" (Gillon 2003, p. 1621). The medical professional is no longer required to (morally) confirm to the norms of the English Gentleman but, rather, to the justificatory approach of modern medical ethics. However this justificatory approach remains situated within the broader culture of medicine and has not been allowed to escape the realities of medical practice. The way of thinking that is modern medical ethics is now an aspect of the modern medical professional's character. Nevertheless, the professional medical persona is greater than being an individual who reflects on the ethical issues of medicine in the prescribed manner even if, in modern times, this is a central aspect of it.

Whilst we can take history of UK medical ethics as an indication that the roots of the contemporary medical profession's willingness to engage in ethical dialogue and debate are connected to the development of medical ethics education at this time this does not provide us with the entire picture. This dialogue and debate was prompted by and, at first, limited to what I would call medical insider–outsiders. Only later did ethical discussions regarding the practice of medicine take place in an manner that included the general public become part (Gillon 1984 suggests this began in the early 1980s) and become enmeshed with wider political discourses (although consider the development of the Great Britian's 1967 Abortion Act which involved some members of the profession). In the UK the development of medical ethics, and medical ethics education looks to have resulted from the activities of a series of medical insider–outsiders with a particular concern for the ethics of medicine it is also something that was cognate with other contemporary changes in medical education. The remainder of this chapter is predominantly focused on elucidating the way in which medical ethics education not only chimes with these changes but was also an integral aspect of their development.

3.2 1993s Tomorrow's Doctors in Context

Given the obvious and continuing influence of the GMC's policy document Tomorrow's Doctors (1993, revised 2003 and 2009) it would be easy to accord the original publication revolutionary status. However such a simplistic interpretation is unsupportable and it is better to see the original Tomorrow's Doctors as one of the final stages in UK medical education's transition, as a field, from a predominantly 'traditional' to a predominantly 'integrated' curricula. This much is clear from the tone of the 1993 document where we find an essentially persuasive and argumentative style. In contrast the 2003 and 2009 revisions offer a more determinate statement.

Despite the impact of Tomorrow's Doctors the GMC's role in medical education remains circumscribed. Whilst it has a responsibility to assess medical schools, their curricula and examinations is has no power to *determine* curricula or to set a core national medical curriculum, nor can it enforce any particular pedagogic approach. However, following the Merrison Report (1975) and the Medical Act (1978) the GMC was awarded the power to co-ordinate all stages of medical education (Irvine 2006, p. 205). Other than the 1980 Recommendations on Basic Medical Education, Tomorrow's Doctors was the first major report concerning undergraduate medical education by the GMC since it had been given this coordinating role. By 1993 changes to postgraduate education had been made and a greater amount of oversight was put in place. Consequentially the need for the medical degree to produce individuals who were in a position to enter any branch of medicine—for medical students be ready to engage in (supervised) medical practice on graduation—was no longer as strong as it once was. Tomorrow's Doctors re-orientated undergraduate medical education towards the production of newly qualified doctors ready to assume the responsibility of a preregistration House officer (currently F1 and F2 doctors) and away from meeting the stipulation that students be in "possession of knowledge and skill requisite for the efficient practice of medicine, surgery, and midwifery" (1993, Sect. 13). Thus the factual content of undergraduate medicine could be safely decreased, as some of the burden could be met during postgraduate education. This freed up space in the perennial battle for curricula time,[8] space into which education of both characterological and a social scientific stripe could be inserted.

It is this sort of curriculum change that 1993s Tomorrow's Doctors is arguing for, particularly in the case of the characterological aspects of the professional role, which were previously considered 'unteachable' (Kaufman et al. 2010, pp. 21–22). Tomorrow's Doctors categorised the goals of medical education as Knowledge and Understanding; Skills; and Attitudes (1993, Sect. 39). The major innovation here is the focus on the attitudinal aspects of medical education and on "student-centered learning [to meet the] ... long-term intellectual and attitudinal demands of professional life" (1993, Sect. 24). This category 'Attitudinal' is not precisely equivalent to what I have been referring to as 'characterological' but this is a major aspect of what is meant. Many of the points (a-l) that fall under the rubric of the Attitudinal Objectives (1993, Sect. 40.2) of medical education require the medical student to develop both personally and professionally and, therefore, require the medical school to aim at this development through the implementation of its curriculum. Communication skills—an essential aspect of the basic clinical method (1993, Sect. 23)—are identified as an explicit theme of the medical curricula (1993, Sect. 46). Finally whilst the GMC's 1980 Recommendations on Basic Medical Education had suggested that instruction in medical ethics should be given (Sect. 64) it was not until the publication of Tomorrow's Doctors that this suggestion became a requirement: point (e) of the Attitudinal Objectives is explicitly concerned with morality and ethics and nearly all the other points in this section have some moral and/or ethical overtones.

---

[8] This concern for ongoing learning is evident in Flexner's early twentieth century survey of medical schools (Calman 2007, p. 230).

Whilst the recommendations that each medical school determine its own core curriculum was (and is) likely to promote the standardisation of medical education it also allows individual medical schools to make virtues of their existing characteristics (cf Brosnan 2008). The recommendation that the core curriculum should be complimented by Special Study Modules (SSMs, later to be renamed Student Selected Components (SSCs)) also allowed for a degree of diversity and provided medical students with the opportunity to pursue their own particular interests at the undergraduate level. As we shall see, this opened up a space within the curricula of UK medical schools that turned out to be a near perfect match for the contemporary drive to expand medical ethics education. As Tomorrows Doctors acknowledges many of the curricula and pedagogic reforms it proposes are, to varying degrees, already being made in UK medical schools. Thus Tomorrow's Doctors was not a top-down revolution of medical education, as Leinster points out some of the "underlying objectives for medical education set out in the document were similar to those in [the Todd report (1968), the Pickering Report (1978)]" (2011, p. 5) and, indeed, the GMC's 1980 recommendations.[9] Tomorrow's Doctors therefore represents another, albeit significant, step in a process of change that had been going on across UK medical school and medical education throughout the 1980s and, given the detail presented in the next chapter, as far back as the 1970s and the introduction of General Practice to undergraduate medical curricula. Nevertheless, its significance should not be underestimated. Tomorrows Doctors set the agenda for medical education in the UK for a generation. Its impact continues to be felt across the (social) field of medical education and within the profession more generally.

One way in which this can be seen is in the professionalisation of medical education. The call for a greater integration across the curriculum led to more centralised control of medical departments by the medical school (Leinster 2011, p. 8). Furthermore a range of non-clinical medical educators was introduced to the medical school, the medical ethicist being only one such example. Such educators were often part of the professionalisation of medical education itself. The Association for the Study of Medical Education (ASME) was formed in 1957 and first published The British Journal of Medical Education (latterly Medical Education) in 1966. Nevertheless it was not until the 1990s that the 'specialist' medical educator (clinical and non-clinical) became common to UK medical schools. The professionalisation of medical education has been identified as an ongoing strategy (Swanwick 2010, pp. xvi and 407) and the exponential increase in research in this area is but one aspect of this.

Whilst medical education did not undergo extensive professionalisation until after Tomorrow's Doctors there were, nevertheless, some signs of this trend prior to its publication, particularly in global context where we can point to the development of PBL. In the case of medical ethics education the early move towards professionalisation is clearer. The detail given in the last section can be

---

[9] Leinster (2011) offers a useful survey of the contemporary influences on Tomorrow's Doctors.

understood not only as an account of the introduction applied ethics into the medical profession and medical education but also as the historical precursor of the 'medical ethicist' in the UK. Of course the connection between medical ethics, medical law and, ultimately, the academic (inter)discipline 'bioethics' has contributed to the production of those medical ethicists who are now situated within UK medical schools. Over the past three decades UK medical schools have increasingly employed specialists in non-clinical subjects, including medical and bio- ethicists, in order to meet the demands of Tomorrows Doctors. Sometimes, particularly in the early part of the development of UK medical ethics education, these individuals have been clinically trained and have a specific interest in (medical) ethics. However in more recent times this has meant employing an ethicist who is 'non-clinical' and whose academic background is in Law, Philosophy, Applied Ethics or, more rarely in the UK, Sociology. Whilst the professionalisation of such ethicists is more clearly linked to the establishment of 'bioethics' than to the professionalisation of medical education per se changes within medical education are clearly relevant to the opportunities that now face medical and bio- ethicists. Nevertheless, the increasing cultural and academic capital of the medical school ethicist is not unrelated to that of their home discipline and the departments and schools and, perhaps more significantly, the centers or institutes that have formally institutionalised bioethics (and related terms such as healthcare ethics, applied ethics, etc.) within the academy. Many, but not all, medical school ethicists remain insider–outsiders.

The institutionalisation of bioethics in the UK has been directly related to the development of medical ethics more generally. However compared to 'medical ethics' bioethics is a relative newcomer to the UK academic scene. This can be most clearly seen in the fact that centres of bioethics were often set up in order to meet an existing demand for medical and professional ethics rather than being preexisting phenomenon that the medical and other healthcare professions could draw upon. Unlike in America where contemporary medical ethics was born alongside and as a part of bioethics (Jonsen 1998) the birth of UK bioethics clearly postdates that of medical ethics and, in a strong sense, has been led by the professions interest in the subject, albeit an interest that has, itself, been led by those who are clearly medical 'insider–outsiders.' It is clear that in the case of medical ethical educators there is a continued trend, perhaps a preference, perhaps a structural imperative, within the medical school to co-opt outsiders and, in so doing, render them insider–outsiders rather than simply outsiders.

From the early 1960s to the late 1980s a broad range of activities were undertaken by these insider–outsiders. Some have been mentioned in the previous section, but we might also include also Cecily Saunders, Alastair Campbell, Robert Downie, Gordon Dunstan, Kenneth Boyd, Tony Hope, Len Doyal, Bill Fulford, Roger Higgs, amongst others.[10] Their broader activities are inseparable

---

[10] For brief biographical details, and others who were involved in the development of UK medical ethics, see Reynolds and Tansey (2007, pp. 169–192).

from the development of UK medical ethics and medical ethics education. Shotter says of Saunders that whilst she consistently lectured under the same title the talk developed significantly over the years. He suggests "it became clear, in retrospect, that she had been reporting the emergence of palliative care as a new specialty... [and that t]here was an interesting symbiosis between the emergence of medical ethics in medical education and of palliative care in medical practice" (Shotter 2006, p. 309). However other than the inception of the LMG it is not easy to point to events that clearly mark out the stages of its development. Certainly one can consider major *bioethical* events such as Kennedy's Reith Lectures (Wilson 2011a), the Warnock Committee (Wilson 2011b) or the landmark 'informed consent' cases Sidaway and Gillick (Gillon 2003, p. 1617) to have been influential, but determining the exact nature of this influence for the purpose at hand, the development of medical ethics education, is problematic; neither were directly concerned with the issue. Although, as Kennedy (1983, pp. 112–118) notes the implications of his thesis are that medical ethics education should be more widespread and systematically approach in undergraduate, postgraduate and continuing professional education. We might think similarly for the influence of the Anglo-American conference (Callahan et al. 1988), Gillon's visit to America, the foundation of the IME (and the JME), and the Pond report. These are all highly relevant to the development of medical ethics education but, nevertheless, determining any exact influence on medical ethics education and, particularly, on the GMC's Tomorrow's Doctors is no easy task.

However many commentators and contemporary protagonists have considered both Tomorrow's Doctors and the Pond Report to be linked by the socio-historical conditions in which they were produced (Reynolds and Tansey 2007, pp. xxiii–xxiv, 59–60) and we might also draw the Oxford Practice Skills Project into this mix. This project, which began in October 1989 and funded by the Leverhulme Trust, produced the widely used text 'The Oxford Practice Skills Course: Ethics, Law, and Communication Skills in Health Care Education' (Hope et al. 1996). Professor David Shaw, the then chair of the GMC's Education Committee the body responsible for the production of Tomorrow's Doctors, was a member of the project's advisory board. A leading project member, Professor Tony Hope, suggests Shaw actively discharged his responsibilities in regards the Oxford practice skills project (personal communication with the author, email dated 17/7/2009). A major innovation in both the Oxford project and Tomorrow's Doctors is the importance accorded to communication skills, medical ethics and law (Hope and Fulford 1994) in medical education and training.

Being a more focused document the Pond Report does not discuss communication skills in any detail but does highlight their importance (Pond 1987, p. 13). Whilst it is tempting to trace the history of modern medical ethics as a specific strand in the recent history of medicine its link to wider changes in the history of medical education should not be neglected. Doing so is a reminder that the practice of professional medical ethics rightly differs from academic or applied philosophical ethics even when focused on similar questions. Thus we can conclude that the pedagogical endeavour of formal ethics is not the same in the classrooms of philosophy and medical students. In the case of the latter it must be linked to the wider processes of the medical apprenticeship, the topic of the final chapter. However in the remainder of this chapter I discuss the nature of this pedagogical endeavour pre and post Tomorrow's Doctors whilst, in the next chapter, providing a specific case study of the development of medical ethics education in the Belfast Medical School, one of the few UK medical schools to not be directly visited by representative of the LMG.

3.3 UK Medical Ethics Education Pre-Tomorrow's Doctors

Throughout the 1980s medical ethics increasingly became part of the undergraduate medical curriculum (Whong-Barr 2003, p. 79). Nevertheless this was largely down to particular individuals with particular concerns and experiences making specific local arrangements as part of courses they already taught. Medical ethics education did not yet proceed on a systematised basis across the 5 years of undergraduate medical education and certainly not beyond undergraduate education.[11] There was not a widespread tradition of offering an introductory foundation and no specialist modules. Whong-Barr laments this lack of ethics-as-specialism suggesting it acted to refocus the concerns of medical ethics away from ethico-political issues prevalent in early LMG lectures towards narrower question of "what to do" (2003, p. 81). However, insofar as specialism can result in curricula isolation it may not necessarily be a wholly positive phenomenon. We might highlight how medical ethics ought to be an endeavour integrated across all aspects of medical education so as to promote the characterological ends of such pedagogy. Understanding its purpose as being in no small part focused on the moral and ethical *development* of medical students as an aspect of their broader professional development questions precisely how important the ethico-political and broader 'critical' aspects of medical ethics education might be.

Thus whilst it is certain that there has been a loss of 'the political' that was part of the initial LMG conception of medical ethics if we place this in the context of developing medical ethics education this may be understandable. Prefacing a JME special issue on the teaching of medical ethics an unattributed editorial (1987), likely by Gillon who talks about this issue elsewhere (Gillan 1990), attends to a neglected distinction in of medical ethics. On the one hand we have traditional medical morality, the embedded norms of the profession reproduced through

---

[11] The SSME was originally set up as a postgraduate study group by 'alumni' of the LMG as it initially proceeded by attempting to circumvent the hierarchical structure of medicine to its own advantage. Fully established medical professional (predominately consultants) and undergraduate students were deliberately brought together as it was felt that, as with all initiates, medical students were accorded a certain freedom to question established medical practices and professionals in a manner that was not open to junior doctors (Whong-Barr 2003, pp. 76–77).

socialisation, and on the other critical medical ethics, modeled on applied philosophical ethics inculcated through formal educational endeavours. Roughly
speaking it is these two conceptions—medical morality and medical ethics—that I
seek to connect in this book. Whilst at its inception modern medical ethics existed
in tension with medical morality—particularly insofar as it sought to modify the
substantive judgements of medical morality[12]—it has now become sufficiently
embedded in medical culture to be considered part of everyday medical morality.
In the early days of modern medical ethics it was still a relatively peripheral
phenomena. At this point modern medical ethics conflicted with the prevalent
medical morality and, in doing so, it was an inherently political, and not simply
ethical, project as it sought to effect change to the moral structure(s) of the profession. The intervening years has seen this new way of (ethical) thinking became
part of medical discourse, practice and education[13]. As this change came about, for
example as with medical ethics becoming more fully (and formally) embedded
within the medical curriculum, something of its political nature was lost. It should
not come as a surprise that the price of this success is that the more radical and/or
esoteric (i.e. less obviously applicable to everyday medical practice) challenges
raised by medico-ethico-political questions would disappear from view.[14] These
are not a matter of any great concern for everyday medical practice. Therefore
neither are they a concern for basic medical education

In its summary of existing arrangements for medical ethics teaching the Pond
Report concludes that whilst there is opportunity for extra-curricula discussion of
ethics (presumably lectures and discussion groups on the model of the LMG) "[m]ost
Deans consider ethics teaching 'important,' [they] are doubtful about introducing it
as a 'separate subject'" (Pond 1987, p. 33). Furthermore few ethics educators had any
specific training in medical ethics or medical ethics education and the amount of
ethics education that actually occurs is highly variable. Nevertheless specific classes,
if not courses or modules, in medical ethics were beginning to take hold (Gillon 1987
and remainder of issue; Irwin et al. 1988; Jones and Metcalfe 1976; Jewell 1984;
Walton 1985 and remainder of issue). The process we see is not simply a response to
an ethical critique resulting in a transformation in substantive medico-moral

---

[12] See, for example how Hafferty and Franks (Hafferty and Franks 1994) contend that formal
ethics education was imagined to be something of a 'magic bullet' capable of immunising
medical students against their exposure to the more negative aspects of medical practice.

[13] Since writing this chapter Reubi (2013) has produced a complimentary perspective which
considers the activities of many of those discussed here in terms of a bioethical 'thought-
collective.

[14] This is not to say that undergraduate medicine should not address what I have called 'medical
politics' (Emmerich 2011b). But I would say that this is not the same a simply reintroducing
ethico-political questions to the medical ethics classroom and I would also argue that whilst it is
important for the medical profession to engage with political and ethico-political discourses this
does not require all medical professionals to be educated to do so. Rather it demands a certain
structure to the profession and its institutions to facilitate such engagement and there being
opportunity for specialist training, in the form of intercalation and masters level degrees, and,
further, giving those with humanities degrees access to the graduate entry medical degree.

judgements. Rather we see an embedding of the new ethical discourse, a new way of thinking, doing and practicing ethics, into the culture of medicine. In no small part, this has been accomplished via medical education. This is best seen restructuring of the moral dimension of medicine as medical ethical reflection becomes a normative aspect of practice. Tomorrow's Doctors marks an important point in this process where medical ethics education went from being largely informal, extra-curricula and dependent on the interest of local medical educators to being a formal requirement of the core medical curriculum. Indeed this can also be seen in the winding up of the LMG in 1989. Whilst financial imperatives were a major factor there was a sense of the job being done as the GMC education committee had been persuaded that ethics could be taught and were set to recommend it be incorporated into the curriculum (Reynolds and Tansey 2007, p. 66 and 104). This also seems to be factor in the closure of at least one other such Medical Group (Reynolds and Tansey 2007, p. 67).

Pre- Tomorrows Doctors medical ethics education was first developed in an extra-curricula context. Subsequently there was gradual incorporation of such activities into the curriculum until, in the mid to late 1980s the first formal under-graduate course stated to appear. For the most part these remained individual classes, or a short series of classes, that delivered in the context of a larger 'module' or, more usually, clinical placements. It was not until Tomorrow's Doctors introduced the notion of the core curriculum and the SSC that medical ethics education became more specialised and, subsequentially, a fully integrated pedagogic endeavour.

3.4 UK Medical Ethics Education Post-Tomorrow's Doctors

Following the publication of Tomorrow's Doctors medical ethics education rapidly became part of most medical schools core curriculum. Whilst an article assessing the implementation of Tomorrows Doctor's based on data collected between 1995 and 1998 makes little direct mention of ethics it finds that most medical schools have reduced the burden of factual knowledge, defined a core curriculum and introduced SSMs (Christopher et al. 2002, p. 284) all key factors in facilitating ethics education. In 1998 the JME published a Consensus Statement for a Core Curriculum in Medical Ethics (revised in 2010). In common with previous statements on medical ethics education and medical education more generally, it consolidates and disseminates practices that some medical schools are already engaged in. It also looks towards the near future in regards making recommendations for the continuation, spread and development of certain approaches to medical ethics education. In undertaking these kinds of activities, and in securing itself an official place within the undergraduate 'core' medical curriculum, we can see the emergence of medical ethics as a kind of speciality. Specialist teachers of ethics are now found in most, if not all, UK medical schools and it is no longer the responsibility of particular individuals who happen to have an interest in the subject.

There are a number of identifiable approaches to formal medical ethics education as currently practiced by medical educators in the UK. At an early point in the degree course, students are introduced to medical ethics in abstract terms. One would expect this to occur in the first year of a medical degree and, although it may take place in the second year, it will be part of the 'pre-clinical' curriculum. The classes will likely take the form of an introduction to applied ethics coupled with a focus on the four principles of Beauchamp and Childress (2008) and various professional codes of ethics. It is also likely that this will be coupled with some case based reasoning with cases drawn either from the applied philosopher's repertoire or from the recent history of medicine. These cases will be robust in the sense that they will resist easy answer from any one ethical perspective. The parameters of the case will be easily manipulated in order to further illuminate not only the ethical problem at hand but the tools, theories, principles, guidelines, laws, and perhaps most importantly the (linguistic) concepts used in any ethical analysis. In short cases that are amenable to metacogntivie examination and manipulation. Theories drawn from moral philosophy may make an appearance, but minimally so, and it is highly likely to be restricted to utilitarian and deontological theories whilst virtue ethics remains conspicuous by its absence.

Whilst co-opting the discourse of applied philosophical ethics medical educators have realised that it must also be appropriately contextualised and integrated into medical education. The 'integration' promoted by Tomorrow's Doctors is an attempt to bridge the structural aspects of medicine such as the preclinical/clinical divide and the medical specialties. Correspondingly integration might be vertical, over the years of medical education, or horizontal, across modules and clinical placements (Christopher et al. 2002). There is a general approach in medical education which involves a move from more classroom based, theoretical, introductions in the early years of a medical degree to a more contextual application and examination. During the clinical years of an undergraduate medical degree the issue of medical ethics is, ideally at least, periodically revisited and although this may be subject to the inclinations of individual clinical educators towards medical ethical discussion most medical students will, in at least some of their clinical placements, be exposed to and engage in reflection on the ethical dimensions of medical practice. One might expect explicit reference to be made to such knowledge, i.e. to the four principles, to certain ethical theory or to particular codes of ethics. Discussions on placement and experiences might be linked to more formal exercises, e.g. the writing up of experiences as an ethics case study. These might then form the basis of further classroom based reflection and analysis, exercises that may include in house ethicists or ethicists from other parts of the university.

Finally more focused medical ethics education is often delivered via the SSCs introduced by Tomorrow's Doctors.[15] Such modules vary in their content and

---

[15] It is worth noting that SSCs in medical ethics are pre-dated by intercalation degrees as well as postgraduate certificates/Master level courses in ethics aimed at healthcare professionals. These

context although there is usually a focus on a medical specialism of particular ethical relevance or on a particular kind of ethical issue raised by modern medical practice. So, for example the Medical School at QUB has offered modules entitled 'Ethics and Paediatrics' and 'Murder and Mercy.' Such courses may be formal and classroom based and includes non-medically qualified individuals including those from beyond the medical school. They may be clinician led and involve time spent on medical wards and experiencing actual practice. Such experiences then form the basis for discussion. There may also be a mixed formal and informal approach.

In 2004 the IME and the BMA commissioned a survey of medical ethics education which was subsequently published in the JME (Mattick and Bligh 2006). The report found that whilst there were feelings of opportunity regarding future development of medical ethics education there was some uncertainty regarding assessment. Nevertheless there is an identifiable common approach: it is primarily delivered through a reflective approach to pedagogy, adopts lectures followed by small group discussion and has a formative (as opposed to summative) mode of assessment. What is perhaps most interesting is that they found that medical ethics was both vertical and horizontal integration across the curriculum i.e. teaching took place in clinical and preclinical education, and in both formal and informal contexts. There is then at least some attempt to relate later teaching to earlier, at least in the planning of medical ethics education. It is then indisputable that in the two decades since Tomorrows Doctors medical ethics has become fully embedded within modern medical education and that it is a fully-fledged aspect of contemporary medical practice.

3.5   Conclusion

I have argued that UK medical ethics and, particularly, UK medical ethics education has roots in concerns for the pastoral care of medical students and the humanist aspects of medical practice and should not simply be understood as the offspring of bioethics, an academic discipline which emerged from the USA. The point of formal ethical education for medical students is not to merely impart knowledge or even argumentative skills but to allow them to explore their own moral concerns and develop their own ethical perspectives. As such it is as much concerned with their broader social, professional and emotional development as it is with the development of analytic skills. Whilst a focus on the cognitive skills of medical ethics has worked to obscure the emotional aspects of medical students development it can, nevertheless, still be perceived in the contemporary focus of medical ethical pedagogy on the exploration, rather than determinate answer, of

---

(Footnote 15 continued)
higher-level qualifications continue to be an important aspect of the field of medical ethics and medical ethics education but are beyond the scope of my discussion.

ethical issues. As such we might understand 'cognitive development' to be an aspect of the broader idea of the medical student's professional development and not simply restricted to the idea of 'analytic skills.' Furthermore whilst the individual patient can slip from view in formal analysis of applied philosophical ethics the integration of medical ethics education into the clinical context seeks to maintain the specificities of the individual patient. At its best it encourages medical students to see the person and their illness and not merely patients and their diseases.[16] Further work is presented in the next chapter to demonstrate that the rise of formal medical ethics education is consistent with and, indeed, apiece of contemporary developments in medical education.

In addition I have presented a case for thinking that as formal medical ethics has become embedded within the medical curriculum and within practice it has come to be part of the broader idea of medical morality. As a particular form of ethics has become normalised within the profession's talk, text and (reflective/metacognitive) practices this has entailed a loss of its critical function. That ethical discourse, and its justificatory purpose, has become commonplace and taken on a kind of regularity, mundanity and sense of the everyday should not be simplistically understood as a criticism but, rather, an indication of the success of the medical ethical project. The critical function of medical ethics can still be located in various activities of the profession—the creation and recreation of ethical guidelines and statements for example—and it can also be found in academic bio- and medical ethical discourses. Such critical ethics is a specialist endeavour. Or, to put it another way, an endeavour that requires medical and non-medical specialists in medical ethics.

These broader discourses are involved in the ongoing project of addressing the ethical aspects of medical practice and, where appropriate, restructuring the institutions of medicine in response. Everyday medical practice takes place within these contexts and is, therefore, structured by them. As evidenced by the existence of medical ethics education as a fully embedded aspect of medical education he long-term project of instituting medical ethics, and medical ethical reflection, as a structure of everyday medical practice has been achieved. Such institutionalisation unavoidable comes with normalisation, regularity, repetition, and a blunting of its critical edge. Or, rather, an appropriate (re)location of the critical edge into the sociological structures rather than into the more mundane clinical practice. A form of medical ethics has become a standardised aspect of medical practice and undergraduate medical education. The second-order ethical practices of medicine have become relatively standardised and, furthermore, aims at producing a relatively standardised product: the nascent medical professional who can talk and think about the ethical issues and cases of medical ethics in the modern, largely principlist, idiom.

---

[16] Mainstream analytic bioethics is, of course, predominantly concerned with patients and diseases. One has to turn to phenomenological, feminist and narrative bioethics to rediscover a concern for individuals and their illnesses.

References

Anon 1998. Consensus Statement: Teaching Medical Ethics and Law Within Medical Education: a Model for the UK Core Curriculum. *Journal of Medical Ethics* 24(3): 188–192.

Beauchamp, T.L., and Childress, J.F. 2008. (First Published 1979). Principles of biomedical ethics, 6th edn. USA: Oxford University Press.

Booth, C. 1994. Obituary: M H Pappworth. *British Medical Journal* 309(6968): 1577–1578.

Boyd, K. 2009. The discourses of bioethics in the United Kingdom. In *The Cambridge world history of medical ethics*, eds. Baker, R.B., and L. McCullough, 486–489. Cambridge: Cambridge University Press.

Brosnan, C. 2008. The sociology of medical education: The struggle for legitimate knowledge in two English medical schools (Unpublished Ph.D. thesis). University of Cambridge.

Callahan, D., Dunstan, G.R., and H. Center. 1988. *Biomedical ethics: an Anglo-American dialogue*. New York: New York Academy of Sciences.

Calman, K.C. 2007. *Medical education*. London: Churchill Livingstone Elsevier.

Christopher, D.F., K. Harte, and C.F. George. 2002. The implementation of tomorrow's doctors. *Medical Education* 36(3): 282–288.

Dunstan, G.R. 1988. Two branches from one stem. In *Biomedical Ethics: an Anglo-American Dialogue*, eds. Callahan, D. G.R. Dunstan, and H. Center, 4–6. New York: New York Academy of Sciences.

Emmerich, N. 2011b. Whatever Happened to Medical Politics? *Journal of Medical Ethics.* 37(10): 631–636.

Gillon, R. 1984. Britain: The public gets involved. *Hastings Center Report* 14(6): 16–17.

Gillon, R. 1986. *Philosophical medical ethics*. Great Britain: Wiley.

Gillon, R. 1987. Medical ethics education. *Journal of Medical Ethics* 13(3): 115–116.

Gillan, R. 1990. Teaching medical ethics: Impressions from the USA. In *Medicine, medical ethics and the value of life*, ed. Byrne, P., 88–115, Chichester: John Wiley & Sons.

Gillon, R. 2003. Section: IV. United Kingdom. Entry: Medical ethics, history of Europe: Contemporary period. In *Encyclopedia of bioethics*, ed. Post, S.G., vol. 3, 3rd edn, 1613–1624.

Gillon, R. 2005. Ploughing a furrow in ethics. *Personal Histories in Health Research,* ed. Oliver, A. 83–97, published in London by the Nuffield Trust. Available from:http://www.nuffieldtrust. org.uk/sites/files/nuffield/publication/personal-histories-in-health-research-aug05.pdf

Hafferty, F.W., and R. Franks. 1994. The hidden curriculum, ethics teaching, and the structure of medical education. *Academic Medicine* 69(11): 861–871.

Hope, T., and K.W. Fulford. 1994. The Oxford practice skills project: Teaching ethics, law and communication skills to clinical medical students. *Journal of Medical Ethics* 20(4): 229–234.

Hope, R.A., K.W.M. Fulford, and A. Yates. 1996. *The Oxford practice skills course: Ethics, law, and communication skills in health care education*. Oxford, UK: Oxford University Press.

Irvine, D. 2006. A short history of the general medical council. *Medical Education* 40(3): 202–211.

Irwin, W.G., R.J. McClelland, R.W. Stout, and M. Stchedroff. 1988. Multidisciplinary teaching in a formal medical ethics course for clinical students. *Journal of Medical Ethics* 14(3): 125–128.

Jewell, M.D. 1984. Teaching medical ethics. *British Medical Journal* 289(6441): 364–365.

Jones, J.P., and D.H. Metcalfe. 1976. The teaching of medical ethics. *Journal of Medical Ethics* 2(2): 83–86.

Jonsen, A.R. 1998. *The birth of bioethics*. New York: Oxford University Press.

Kaufman, D.M., K.V. Mann, and P.A. Jennett. 2010. Teaching and learning in medical education: How theory can inform practice. In *Understanding medical education: evidence, theory and practice*, ed. Swanwick, T., 16–36, UK: ASME and WIley.

Kennedy, I. 1983. (First Published 1981). The unmasking of medicine. Revised ed. London: Flamingo.

Leinster, S.J. 2011 The history of change in the UK. In *The Changing Face of Medical Education*, eds. Cavenagh, P., S.J. Leinster, and S. Miles, 1–12, Uk: Radcliffe Publishing.

Mattick, K., and J. Bligh. 2006. Teaching and assessing medical ethics: Where are we now? *Journal of Medical Ethics* 32(3): 181–185.

Mepham, C.A. 1963. The implications of medical study: A report commissioned by the student christian movement of great Britain and Ireland. *Student World* 2: 157–161.

Merrison, A.W. 1975. *Report of the committee of inquiry into the regulation of the medical profession*. London: H.M.S.O.

Pond, D. 1987. *IME report of a working party on the teaching of medical ethics (the Pond report)*. London: The Institute of Medical Ethics.

Reubi, D. 2013. Re-moralising Medicine: The Bioethical Thought Collective and the Regulation of the Body in British Medical Research. *Social Theory and Health* 11(2): 215–235.

Reynolds, L.A.,and E.M. Tansey. (eds) 2007. *Medical ethics education in britain: 1963–1993*. London: Wellcome Trust Centre for the History of Medicine at UCL.

Shotter, E. 1988. *Twenty-five years of medical ethics*. London: The Institute of Medical Ethics.

Shotter, E.F. 2006. Dame Cicely Saunders. *Journal of Medical Ethics* 32(5): 309.

Stirrat, G.M., C. Johnston, R. Gillon, and K. Boyd. 2010. Medical ethics and law for doctors of tomorrow: The 1998 consensus statement updated. *Journal of Medical Ethics* 36(1): 55–60.

Swanwick, T. (ed.) 2010. *Understanding medical education: Evidence, theory and practice*. Sussex: ASME and Wiley.

Walton, J. 1985. The general medical council's medical ethics education conference. *Journal of Medical Ethics* 11(1): 5.

Whong-Barr, M. 2003. Clinical Ethics Teaching in Britain: a History of the London Medical Group. *New Review of Bioethics* 1(1): 73–84.

Wilson, D. 2011a. Who guards the guardians? Ian Kennedy, bioethics and the 'Ideology of Accountability' in British medicine. *Social History of Medicine*. Forthcoming, Early View. http://shm.oxfordjournals.org/content/early/2011/07/01/shm.hkr090.abstract.

Wilson, D. 2011b. Creating the 'Ethics Industry': Mary Warnock, in vitro fertilization, and the history of bioethics in Britain. *BioSocieties* 6(2): 121–141.

Chapter 4
Prof. W. G. Irwin: A Case Study in the Development of Medical Ethics Education in the UK

4.1 Introduction

On his appointment in 1971 William George Irwin became the UK's second Professor of General Practice and the first on the island of Ireland. Given the high percentage, both then and now, of medical graduates who go on to become GPs it is somewhat startling to note that it was only at this time that medical schools began to directly address General Practice in their curricula. Given the social organisation of medical education the introduction of a Professor of General Practice meant the introduction of a Department of General Practice and so the guarantee of curriculum time. In the context of the ever-increasing drive towards medical specialism, kick-started by the various reports of Abraham Flexnor and having their basis in biomedical science, General Practice, and the other 'Cinderella' subjects of medicine, were at a distinct disadvantage in terms of the kind of knowledge they had to offer. On his appointment Prof WG Irwin's first task was to articulate General Practice as a medical specialism and, in so doing, produce a curriculum and, as we shall see, pedagogy that could meet its specific needs. He also articulated a research program that would distinctively identify General Practice as not only a medical specialism but as an academic area of enquiry. The way in which Irwin accomplished this task is discussed in the first section of this chapter. In the subsequent sections I trace his teaching career from his first foray into medical ethics, a course addressing issues in terminal care, and, via a seat on the Warnock Committee, to the design and implementation of a multi-disciplinary course in medical ethics.

In the last chapter I discussed the history of medical ethics education in the UK and highlighted the role of the London Medical Group and its various off shoots, including other such 'medical groups' and, ultimately, the IME and the JME. This perspective counters the presumption that medical ethics education in the UK resulted from the external development of Bioethics, largely in America. However what is particularly interesting about Irwin's career in the Belfast Medical School is that his contact with the LMG was minimal. Certainly by the 1980s was aware of the JME and the IME, including its Pond Report (1987), nevertheless, unlike

N. Emmerich, *Medical Ethics Education: An Interdisciplinary and Social Theoretical Perspective*, SpringerBriefs in Ethics, DOI: 10.1007/978-3-319-00485-3_4, © The Author(s) 2013

other medical schools in the UK, representative from the LMG never visited Belfast and no comparable group was ever set up. Thus the way in which Irwin and his colleagues approached medical ethics education can be understood as related to his previous innovations in medical education rather than as the direct result of external influences.

This case study counters the idea that medical ethics education in the medical curriculum resulted purely from the activities of the LMG and developments that can be subsequently traced to it. Rather it is one aspect of the wider changes that were occurring in undergraduate medical education at this time. I do not wish to underplay the relevance of the development of bio and medical ethics, both in America and the UK, nor the importance of the LMG. However it is essential not to see medical ethics education and its pedagogical form as, simply, the natural result of the development of a broader concern for medical ethics either in the profession, in other areas of academia, in wider public discourses, or within the LMG. A historical perspective on the development of medical ethics education should acknowledge that it is not only the result of the interplay between bioethics and bioethical discourses, including those found within the LMG and the medical profession, but it is also connected to discourses within medical education. I argue that those that can be traced to the development of General Practice as an academic discipline are of particular importance. The career of Prof Irwin provides us with a case study to support this position.

4.2 Prof. W. G. Irwin's Development of Academic General Practice

Prior to his appointment Professor Irwin taught general practice under the aegis of Prof John Pemberton—another medical pioneer in education and public health (Nicholl and Holland 2010). Irwin was also in General Practice and, further to his primary medical qualification, he had obtained an M.D. (Irwin 1969). The broader history of General Practice education in NI has been masterfully analyzed by Harland (2001, 2003 and 2006) and I am indebted to his work. He suggests that Prof Irwin "ventured into the realms of educational theory in an unprecedented way" (Harland 2003, p. 145) and "developed new methods of teaching, particularly in the field of communication skills [taking advantage of] ... modern teaching aids including CCTV" (Harland 2006, pp. 145–146).

Harland gives some further detail on Irwin's approach to teaching general practice but, as he is more concerned with issues of power and the advent and initial development of academic general practice in NI, he does not fully explore Irwin's pedagogic activities. However he does say that "in 1971 the Medical Faculty had completely remodeled its teaching into an integrated teaching course" (Harland 2001, p. 11). Having been in development for some time, this was the year that, the Department of General Practice was inaugurated and Prof Irwin took up his appointment.

In Irwin (1973) Irwin delivered his inaugural lecture entitled "The Need for Academic General Practice". In it he covers the reasons for considering general practice a medical specialism and sets forth his plans for the newly formed Department. He discusses the academic content of general practice in terms of both teaching and research. The lecture gives us a clear idea of what Irwin envisaged as constituting the general practice curriculum and the aims and objectives he saw as being its driving purpose. Irwin bases his perspective on general practice as a speciality on the quality and depth of the relationship between GP's and their patients arguing that in general practice "the degree of rapport, understanding and communication established during a consultation dictates the success or failure of care" (Irwin 1973, p. 11). Citing Balint's "The Doctor, his Patient and the Illness" (Balint 2000) as a major influence Irwin suggests that the doctor-patient relationship found in general practice is "different, more continuous and intimate" (Irwin 1973, p. 11) than that found within hospital medicine. In doing so he seeks to "indicate the importance attached to the hidden content of many consultations in general practice" (Irwin 1973, p. 7) concluding that:

> Students have to be taught communication skills, how to establish rapport, how to obtain information indirectly by enhanced perception and by cultivating the art of listening. They have to learn the basic skills of primary diagnosis, how to differentiate unselected primary illness with sometimes inadequate evidence and minimal diagnostic resources. They should be taught the responsibilities of decision making and management in primary care (Irwin 1973, p. 11).

Here a qualitative difference between general practice and hospital based medical specialities is being drawn. There is a different ethical quality to the doctor-patient relationship in general practice as compared to hospital medicine precisely because the patient is seen in their social context over time. This forms the basis of a deeper (medical) responsibility on the part of GPs. This fact places the interpersonal, in the form of teaching for communication skills, at the forefront of education in general practice and, furthermore, it also locates general practice at the head of developments in medical education that occur throughout the 1980 and 1990s. In general practice the focus is on the beginning and end stages of a disease or illness, and on the management of chronic illness i.e. outside of the acute stage most often encountered in the hospital. This focus on illness, and the subjective state of the patient, is to be contrasted with disease and the objective state of the patient's body. Thus communication skills and enhanced perception are the specific qualities or skills require by the specialised and expert GP; technical skill and medical knowledge, encyclopaedic or otherwise, will simply not suffice.

The required qualities, such as interpersonal skills, are characterological and not merely epistemic. General practice requires not only the accumulation of technical and scientific knowledge and the rational and cognitive ability to use such knowledge but also the development of the empathic and humanistic self. Irwin's training and vision for general practice is aimed at personal and professional self-development i.e. at an emotional and affective, as well as cognitive, development. This articulation of general practice as a specialty challenged traditional approaches to medical education and, in doing so, indicated an area for research. Although Irwin clearly recognises that alternative educational approaches were required to teach general practice effectively research on such education gets only a small mention towards the end of the inaugural lecture. Nevertheless publishing articles on medical education proved to be a not insubstantial aspect of Irwin's future activities.

The resistance to Irwin's pedagogical innovations should not be under estimated. In 1975 Irwin's colleague, Dr Smiley, give the Annual Oration at the Opening of the Teaching Session at the Royal Victoria Hospital Belfast where he suggested that "[t]here are many things that cannot be taught-other than by example. Lectures and even ward rounds cannot teach you the small courtesies or rapport with the patients" (Smiley 1975, p. 36). Smiley explicitly mentions Irwin's appointment saying that it gives him great pleasure and that "[w]e wish him well, but hope that he will not lecture too much and will remember the words of Osler that 'a true knowledge of medicine is learned at the bedside'" (Smiley 1975, p 36). It is uncertain if Smiley would have recognised general practice placements as an opportunity to 'learn at the bedside' or not. Regardless Irwin did not merely lecture. By the time of Smiley's oration he was already engaged in small group teaching. When Irwin began to research medical education it is notably that this was conducted with one Dr Bamber, of the School of Education, QUB. Bamber is described as a "consultant educationalist" (Irwin et al. 1976, p. 302) and their academic partnership continued for a decade during which time they worked on a diverse range of medical education innovations including: an examination of the characteristics of medical students (Irwin and Bamber 1978a); evaluation of undergraduate general practice courses (Irwin and Bamber 1978b); and an assessment the cognitive structure of the modified essay question (Irwin and Bamber 1982).

Bamber also publishes with Irwin's colleague Prof Stout (Stout and Bamber 1979) with whom Irwin assembled and implemented an integrated curriculum for general practice, community medicine, geriatrics, and mental health (Stout and Irwin 1982). The course covers: communication; terminal care; bereavement; alcohol problems; maternal and child care; the confused elderly patient; disability; human sexuality; coronary care; and screening. Although ethical discussion is not an explicit objective of the class these are all topics with a moral dimensions in that they are all concerned with "the management of the 'whole' patient in his environment... [and aim to show students] medical care in its widest sense" (Stout and Irwin 1982, p. 146). Interestingly Bamber is not the only 'medical outsider' involved with this course. A particularly innovative feature of the course is that it includes clergymen, paramedics, a marriage guidance counselor and, for the session on disability, a paraplegic clinical psychologist in pedagogic roles. Inviting medical outsiders into the classroom to engage the students in discussion and to provide them with a broader point of view for the purposes of reflection is a remarkable innovation. The use of video is also mentioned: students watch scenarios, discuss, report back and rediscuss. In a later paper, focused on the teaching of communication skills, they suggest that the use of CCTV "on a mass scale has been very successful in teaching basic clinical skills to fourth year medical students. These include the ability to think clearly and critically about the process of the consultation in general practice" (Irwin and Bamber 1984). Whilst not termed as such Irwin and colleagues are engaged in the use of reflection as a pedagogic tool for the creation of what (Schön 1984) calls the reflective practitioner.

Whilst Bamber and Irwin are concerned with interweaving interpersonal and empathic communication with clinical skills they draw a theoretically informed "distinction between communication skills acquired through the normal processes of social learning as a member of various social groupings and more technical and formal aspects of communication skills which are appropriate in the professional context but which might not be employed in normal social intercourse" (Irwin and Bamber 1984, p. 91). The importance of communication skills and the relevance of the professional context where the demands of social interactions are 'beyond the normal' are highlighted by the topic of terminal care (Irwin 1984a, b). Whilst we can appreciate that professional communication skills are suffused with moral implications it is only with a more focused concern with terminal care that the ethical aspects of medical practice begin to come to the fore of Irwin's educational activities. We might suppose that this nascent interest is later bolstered by Irwin's participation in the Warnock committee and, taken together, the result is Irwin (1987) setting out his views on the subject and the formal teaching of medical ethics. However what is of primary importance to this case study is Irwin's background as an innovative medical educator, set out above. Many of the innovations discussed—the inclusion of medical outsiders, interdisciplinary approach to undergraduate medical education and the use of reflection—will, as we shall see, recur in his teaching of medical ethics.

4.3 Teaching Terminal Care: A First Foray into the Ethics Of Medicine

The first time Prof Irwin can be seen as having a concern for the ethical aspects of medicine is when he begins teaching terminal care (Irwin 1984a, b). This activity was likely prompted, and certainly informed, by Dr Campbell Moreland's reflections on being a patient. Dr Moreland was a graduate of QUB and in general practice when he was diagnosed with testicular cancer. His initial diagnosis and treatment took place in Belfast, no doubt at a hospital where he had trained. For a time his treatment was transferred to London but, partially due to the demands of travel, this was soon discontinued. Some of the reasons Dr Moreland had for preferring to be cared for in London were directly related to the fact he had trained in Belfast and so with those who were treating him. Dr Moreland published his reflections on being a patient with a terminal diagnosis in an article published by the Lancet (Moreland 1982). An extended version was informally circulating around the Belfast medical school and later published as a pamphlet by QUB (Moreland 1984). The explicit purpose was its use within medical education.

Dr Moreland's reflections offer an insight into the role and phenomenology of being a 'patient' and their relationships with doctors. In his writing Moreland complains that despite being medically qualified, and despite his doctors being aware he was medically qualified, he is not given the full details of his condition and so it cannot be fully discussed. This complaint is common to the discourse of medical ethics at this time which was seeking to ensure that medical professionals informed their patients in order that they could properly consent to treatment. However Dr Moreland's concerns go deeper than simply being informed about his diagnosis, prognosis and treatment. What Dr Moreland is concerned about is the ability of medical professionals to engage with him about his diagnosis and, specifically, the possibility that his is a terminal prognosis. He is concerned about the medical professionals ability meet the needs of (potentially) dying patients and talk about death.

In presenting his experiences Dr Moreland is honest about what it is like to be a patient with a life-threatening illness. Evidently, as it was published in the Lancet, Dr Moreland received a kind of attention that other non-professionals, whether they be patients or other medical interlocutors, would not. Furthermore his view does not 'judge' medicine he is clear that medical professionals are at the mercy of the patient's fluctuating emotional state and that there is little possibility of them being able to simply 'do the right thing.' Nevertheless as a patient he clearly articulates a need that medical professionals are unable to meet. That need is to acknowledge and talk about death when it is a potential consequence of the diagnosis. In his attempt to teach terminal care Irwin clear sets out to meet this need saying that the "whole emphasis of teaching in these sessions is to make students sensitive to the needs of the dying" (Irwin 1984a, p. 1510). His view is that medical students, and by extension medical professionals, are ill-equipped to meet these needs until they become "aware of their own attitudes to death" (Irwin 1984a, p. 1510).

Irwin's course is designed to help students overcome their fears so that they can professionally care for terminally ill patients. Once again both video recordings of general practice consultations are used as a basis of discussion and some medical outsiders, predominantly Christian ministers of various denominations, are invited to take part. However, recalling Balint, the emphasis is clearly on an almost psychodynamic process which facilitates the emotional development of medical students. The mode through which this is accomplished is discussion and, most importantly, reflection, both intellectual and emotional. Through encouraging medical students to examine their own attitudes towards death the course facilitated their engagement with their own 'emotional baggage.' For the purposes of professional development the medical students is required to undergo a form of personal development. The medical student is not required to become emotionally neutral but is required to be able to put themselves to one side to meet the needs of medical practice and maintain a professional demeanor when dealing with dying patients. Irwin's teaching of terminal care represents a commitment on the part of doctors to honesty and respect in their relationships with patients coupled with an understanding that patients are entitled to and 'own' their own emotional lives. Of course, as individuals, medical professionals are also entitled to their own emotional lives. However the professional must learn to manage their emotions in order that they might practice effectively.

4.4 Prof. W. G. Irwin and the Warnock Committee

In the early 1980s Prof Irwin was appointed to the Committee of Inquiry into Human Fertilisation and Embryology, colloquially known as the Warnock Committee. Prof Irwin has suggested that he did not know why he was selected for the committee and that the first he knew about it was when he received a telephone call from Dr Wier of the DHSSNI informing him of his appointment (personal communication, letter to the author 21/04/10). Whilst there has been some recent historical research into the committee (Wilson 2011a) the reasons for Irwin's appointment remain a matter of speculation (Emmerich 2011a, b, c, p. 234). Regardless, Irwin must have been exposed to a range of contemporary thinking on medical and bio- ethics as a result of his participation on the committee. This must have had an impact on his later activities regarding medical ethics and medical ethics education.

The ethical issues considered by the Warnock Committee were largely those raised by biomedical research and the use of human ova, sperm and embryos for fertility treatment. The committee's orientation was towards constructing a workable framework to facilitate the developing area of embryological research (Warnock 1985) and so was more 'bioethical' than 'medical ethical'. Whilst it would be incorrect to maintain a hard distinction between bioethics in the committee room and professional medical ethics at the bedside we can, nevertheless, sketch a meaningful differentiation, particularly with regard to medical education (Rhodes 2002).

In the context of Prof Irwin's previous excursions in medical education we might suppose that this exposure to medical ethics simply prompted him to teach medical ethics i.e. that medical ethics was just another topic that could be introduced to the curriculum. However, this would be overly simplistic. There is a deeper connection. As with communication skills and character education, ethics was often held to be something that could not be taught and could only be learnt by osmosis at the bedside or selected for on entrance to medical school. Irwin's previous experience in teaching such unteachable aspects of medical practice, and not just his involvement in medical education, is obviously important to his subsequent teaching of medical ethics.

Furthermore, as recognised by a JME editorial in 1985, the successful teaching of medical ethics requires the input of medical outsiders, something Irwin was open to. Medical ethics education requires the frank discussion of difficult topics with the aim being the development of a balanced and professional perspective on the ethics of medical practice on the part of the medical students. There is an obvious 'reflective' aspect to medical ethics education. Whilst academic bioethicists are, perhaps, predisposed to see this ethical reflection and medical ethics education as an exercise in rational and critical thinking the affective and emotional aspects of this process cannot, and should not, be underestimated. Given his previous activities Irwin was well placed to fully comprehend the nature and purpose of medical ethics curricula within medical education.

4.5 The Development of Medical Ethics Education by Prof. W. G. Irwin and Colleagues

The Presidential Addresses given in 1981 and 1986 to the Ulster Medical Society concerned medical ethics. The first, given by a surgeon Willoughby (Wilson 1982), was titled 'Old Ethics: New Dilemmas' whilst the second, given by Irwin (1987), was simply called 'Medical Ethics.' Taken together they mark the transition from the old ethics to the new on the part of the Belfast Medical Profession. Where Wilson advances a paternalist, etiquette based ethics based on the "the six well known 'A's" (1982, p. 27)[1] and calls on all medical students to be 'gentlemen' (1982, p. 28)[2] Irwin offers a resolutely modern, principlist and 'bioethical' account. Whilst it is clear that Irwin has been influenced by the famous four principles (Beauchamp and Childress 2008; Gillon 1986) it is also clear that he has given the topic a good deal of thought. Irwin articulates seven principles of medical ethics transmuting autonomy into 'respect for the authority of the patient' and adding truth telling, preservation of life and confidentiality to the well-known beneficence non- maleficence and justice (Irwin 1987).

Irwin's promulgation of seven, rather than four, principles of medical ethics demonstrates the way in which his concerns, as a medical professional and educator, differ from those of Beauchamp and Childress, as philosophers and theorists of applied ethics in the domain of medical research and practice. For example Irwin places medical practice and experience in a primary position. There seems little point to a medical ethics that is fully independent of medical practice. Whilst it is obvious that there is no morally neutral medicine (Canguilhem 1991) it is also true that there is no ethically neutral medicine; whether implicit or explicit, recognised or not, all medical practice has an ethical dimension. If, at a deep level,

---

[1] These are: Abortion; Adultery; Alcohol; Addiction; Association; and Advertising. Despite Wilson's claim that they are 'well known' I have founded little reference to them other than the entry 'A's, Rule of.' in The New Dictionary of Medical Ethics (Boyd et al. 1997).

[2] Including, one presumes, the female medical students who, he acknowledges, constitute one-third of the cohort (Wilson 1982, p. 29).

medical practice and medical ethics are unavoidably related to one another then philosophical ethical reasoning alone cannot solve the moral problems of medicine. Thus Irwin warns that too much analysis of morality might threaten the ethical project itself suggesting one ought not scrutinise the "foundations of morality [too hard lest] the superstructure will come tumbling down" (Irwin 1987, p. 1).[3]

Medical doctors are not, and cannot be, moral philosophers or even applied (bio) ethicists. The demands of practice, the demands for action, are such that ethical reflection cannot be permitted to dominate proceedings and risk professional paralysis. Irwin perceives medical ethics and medical practice to be mutually constitutive and so his overarching aim is "to try and express effectively an acceptable and practical framework of ethical principles which could provide a basis of moral reasoning in medical practice" (Irwin 1987, p. 1). We might also add that this framework it is not philosophical concise and has no need to be. It is designed to meet the needs of medical education and the demands of future medical practice.

In the final section of his address Irwin discusses the teaching of medical ethics. In a passage addressing similar concerns that have been raised regarding the teaching of communication skills Irwin notes that the by "the time students reach medical school, their moral character has been formed" (Irwin 1987, p. 11). Nevertheless medical education can still equip students with "ethical knowledge and interpersonal skills to enhance their ethical behaviour" (Irwin 1987, p. 11). As a medical educator there is, for Irwin, a pedagogic continuity between communication skills and medical ethics and it is around this time that he acts as an Academic and Clinical Advisor to Dr (later Professor) Len Doyal's project which created the Nuffield Video Library in Medical Ethics and Law, published in 1990. Some of the scripting and filiming of the footage in these videos was done in the Television Studio attached to the Department of General Practice in Belfast and Irwin appears as the discussant in episode 6 entitled 'The Moral Problems concerning informed consent which typically confront General Practitioners.'

In 1988 Irwin and colleagues publish a paper in the JME detailing their multidisciplinary teaching of formal medical ethics to clinical students (Irwin et al. 1988). All sessions are based around particular cases and the focus is on discussion. Irwin's openness towards medical outsiders in medical education is again on display as the classes include various members of QUB staff including philosophers, a member of the law school as well as those from further afield including someone from the DHSSNI. The first section of the paper details learning objectives for the course, perhaps indicating the collective and individual experience and approach to medical education preferred by the authors. Reference is made to the GMC's 1980 commitment to the teaching of medical ethics (General Medical Council: Education Committee 1980), the IME's 'Pond Report' (1987), Irwin's Presidential address 'Medical Ethics' (Irwin 1987), and a previous

---

[3] Irwin cites Samuel Butler as the source of this passage but gives no further reference.

publications regarding integrated medical education (Stout and Irwin 1982). Most interesting however is reference to the work 'Human Values in Health Care' (Wright 1987) which underpins the course objectives, these being:

(1) To help the student to analyse competently a clinical situation or problem;
(2) To help the student to identify moral issues inherent in clinical situations;
(3) To introduce a range of moral concepts, principles and codes used frequently in the discussions of medical ethics and to relate them to everyday morality;
(4) To encourage the student to examine his or her own values, beliefs and attitudes and to relate to those of others;
(5) To encourage the student to give his own views, based on reasoned moral argument, on moral issues related to practice (Irwin et al. 1988).

It seems to me that these objectives might be reordered and better be expressed as: The aim of the course is to introduce a range of moral concepts, principles and codes used frequently in the discussions of medical ethics and to relate them to everyday morality (3). This is done in order that the student might identify moral issues inherent in clinical situations (1) and competently analyse the ethical dimensions of a clinical situation or problem (2). The wider aim is that the student is encouraged to give (and develop) their own views on moral issues related to practice, based on reasoned moral argument (5). In so doing students are encouraged to examine their own values, beliefs and attitudes and to relate to those of others (4).

This is very much a 'student centered' approach to medical ethics education and in line with the nascent developments in medical education at this time and later promulgated by Tomorrow's Doctors. Prof Irwin has suggested to me that in his "experience of teaching Medical Ethics to medical students [it] must be patient centered if they are to benefit from it" (Personal Communication dated 13/4/2010). What I understand Prof Irwin to mean is that students must be able to appreciate not only the clinical relevance of medical ethics education but also that medical ethics, the ethics of doctors, ought to be founded on the interests of the patient.[4] This perspective also provides some nuance to the idea of student centered education. As with the case of teaching terminal care, the medical student is being encouraged to reflect and develop their own ethical perspective on medicine so that they are prepared for professional practice. It is detrimental to the patient's well-being and interests if medical professionals are not adequately equipped to engage with the ethical issues that will, inevitably, arise. Medical students are not being asked to work out or finalise their own ethical opinions such that they cover all possible eventualities but they are being asked to reflect on their own ethical opinions and those appropriate to the profession. In part the aim is to ensure that their own need to grapple with the moral and ethical features of life and medicine

---

[4] If this is the case than I think Prof Irwin is, like Ranaan Gillion, suggesting that whilst all the principles of medical ethics are of equal importance it is nevertheless the case that one is more equal than the others. Unlike Gillion (2003), and like Pellegrino (1989), I think Irwin would say that this principle is beneficence and not autonomy/respect for the authority of the patient.

at a personal level does not interfere with their moral and ethical responsibilities when they commence medical practice. The aim is also to provide medical students with a set of common, secular, concepts with which they can address the ethical issues of medical practice as professionals. In teaching terminal care the aim was for medical student to confront their own emotional responses to death and dying, in teaching ethics the aim is not dissimilar—students are encouraged to confront their own ethical responses to various aspects of medical practice. However there is also an engagement with the rational and reflective aspects of modern medical ethics as this is now a central aspect of modern medicine. Medical professionals must be able to give ethical reasons for their ethical actions.

4.6   Conclusion

Both medical ethics and communication skills are examples of 'soft skills' that were previously considered unteachable. The gradual acceptance of each is representative of the modernisation of medical education in the UK and the slow transition from one era to another. Irwin's early, and pioneering, adoption of these topics in his general practice curricula is indicative of the place of general practice in the modernisation of medical education. Irwin's innovation in teaching medical ethics may be unusual and related to his experiences regarding terminal care, Dr Campbell Moreland and the Warnock Committee rather than to academic general practice. Nevertheless there is an important connection between this and the teaching of communication skills, particularly regarding their shared characterological dimensions. Having surveyed the advent of medical ethics education in the UK and in the QUB medical school I conclude that it cannot be adequately understood outside of changes within medicine, medical culture and medical education more generally. Trained philosophers seem to be relative latecomers to medical ethics education even if they are currently predominant in the discipline of bioethics and, alongside colleagues from law, appear to provide the majority of advanced (Masters) level education available in the UK, including intercalation degrees. Nevertheless the teaching of medical ethics as a part of a professional and vocation medical degree must be understood in context and not be mistaken for a 'transplanted' philosophy or applied ethics class. Developments subsequent to those discussed in this chapter have seen medical ethics education integrated into clinical placements and the added dimension of experience and its use within more formal reflection indicates that medical socialisation and education in ethics must be understood as being connected. This is particularly true in so far as reflective forms of education, and ethics, are utilised and enculturated.

References

Balint, M. 2000. *The doctor, his patient and the illness*, 2nd ed. Philadelphia: Churchill Livingstone.

Beauchamp, T.L., and J.F. Childress. 2008. *(First Published 1979). Principles of biomedical ethics*, 6th ed. USA: Oxford University Press.

Boyd, K.M., R. Higgs, and A. Pinching. 1997. *The new dictionary of medical ethics*. London: BMJ Publishing Group.

Canguilhem, G. 1991. *The normal and the pathological*, New ed. New York: Zone Books.

Emmerich, N., 2011a. *Taking education seriously: Developing bourdieuan social theory in the context of teaching and learning medical ethics in the UK undergraduate medical degree.* (Unpublished Ph.D Thesis). Belfast: Queen's University Belfast.

Emmerich, N. 2011b. Whatever happened to medical politics? *Journal of Medical Ethics* 37(10): 631–636.

Emmerich, N. 2011c. Literature, history and the humanization of bioethics. *Bioethics* 25(1): 112–118.

Gillon, R. 1986. *Philosophical medical ethics*. Great Britain: Wiley.

Gillon, R., 2003. Section: IV. United Kingdom. Entry: Medical ethics, history of europe: Contemporary period, 3rd edition, vol 3. In *Encyclopedia of bioethics,* ed. Post, S.G, 1613–1624.

Harland, R. 2001. The history of the teaching of the specialty of general practice in northern Ireland. (Presidential address to the ulster medical society). *The Ulster Medical Journal* 70(1): 5–14.

Harland, R.W., 2003. The history of the teaching of the specialty of general practice in Northern Ireland (Unpublished Ph.D Thesis). Belfast: Queen's University Belfast.

Harland, R.W. 2006. GP education in Northern Ireland 1920–1990: A Study of the use and misuse of power (The 2005 rose prize essay). *The Ulster Medical Journal* 75(2): 141–152.

Irwin, W.G., 1969. Multiple screening with the technicon autoanalyzer in a general practice at finaghy health centre, Co. Antrim (Unpublished MD Thesis). Belfast: The Queens' University of Belfast.

Irwin, W.G. 1973. *The need for academic general practice: An inaugural lecture*. Belfast: Queen's University of Belfast.

Irwin, W.G. 1984a. Teaching terminal care at Queen's University of Belfast I: Course, sessional educational objectives, and content. *British Medical Journal* 289(6457): 1509–1511.

Irwin, W.G. 1984b. Teaching terminal care at Queen's University of Belfast II: Teaching arrangements and assessment of topic. *British Medical Journal* 289(6458): 1604–1605.

Irwin, W.G. 1987. Medical ethics (Presidential address to the ulster medical society). *The Ulster Medical Journal* 56(1): 1–12.

Irwin, W.G., and J.H. Bamber. 1978a. An Evaluation of a Course for Undergraduate Teaching of General Practice. *Medical Education* 12(1): 20–25.

Irwin, W.G., and J.H. Bamber. 1978b. Characteristics of Senior Medical Students at Belfast. *Medical Education* 12(2): 117–123.

Irwin, W.G., and J.H. Bamber. 1982. The Cognitive Structure of the Modified Essay Question. *Medical Education* 16(6): 326–331.

Irwin, W.G., and J.H. Bamber. 1984. An Evaluation of Medical Student Behaviours in Communication. *Medical Education* 18(2): 90–95.

Irwin, W.G., J.H. Bamber, and J. Henneman. 1976. Constructing a New Course for Undergraduate Teaching of General Practice. *Medical Education* 10(4): 302–308.

Irwin, W.G., R.J. McClelland, R.W. Stout, and M. Stchedroff. 1988. Multidisciplinary teaching in a formal medical ethics course for clinical students. *Journal of Medical Ethics* 14(3): 125–128.

Moreland, C. 1982. Disabilities and how to live with them: Teratoma of the testis. *Lancet* 318(8291): 203–205.

Moreland, C. 1984. *Whose choice? Whose consent?*. Belfast: Queen's University Belfast.

Nicholl, J., and W. Holland. 2010. John Pemberton. *British Medical Journal* 340(7753): 977.

Pellegrino, Edmund D. Character, Virtue and Self-Interest in the Ethics of the Professions. *Journal of Contemporary Health Law and Policy* 5: 53–73.

Pond, D. 1987. *IME report of a working party on the teaching of medical ethics (The pond report)*. London: The Institute of Medical Ethics.

Rhodes, R. 2002. Two concepts of medical ethics and their implications for medical ethics education. *Journal of Medicine and Philosophy* 27(4): 493–508.

Schön, D.A. 1984. *The reflective practitioner: How professionals think in action*. USA: Basic Books.

Smiley, T. B. 1975. Medical Students and Their Education. *The Ulster Medical Journal* 44(1): 28–38.

Stout, R.W., and J.H. Bamber. 1979. A New Undergraduate Teaching Course in Geriatric Medicine. *Medical Education* 13(5): 363–367.

Stout, R. W., and W.G. Irwin. 1982. Integrated Medical Student Teaching. A Combined Course in Community Medicine, General Practice, Geriatric Medicine and Mental Health. *Medical Education* 16(3): 143–146.

Warnock, M. 1985. Moral thinking and government policy: The warnock committee on human embryology. *The Milbank Memorial Fund Quarterly* 63(3): 504–522.

Wilson, W. 1982. Old ethics: New dilemmas. *The Ulster Medical Journal* 51(1): 23–34.

Wilson, D., 2011a. Who guards the guardians? Ian Kennedy, bioethics and the 'ideology of accountability' in British medicine. *Social history of medicine*. Forthcoming, Early View. http://shm.oxfordjournals.org/content/early/2011/07/01/shm.hkr090.abstract.

Wright, Richard A. 1987. *Human Values in Health Care: The Practice of Ethics*. New York, USA: McGraw-Hill.

Chapter 5
Medical Ethics Education from a Socio-Cultural Perspective

5.1 Introduction

As I have discussed drawing on the insights offered by psychology is unavoidable in any attempt to articulate an interdisciplinary perspective on professional reproduction that encompasses and connects the 'social' and 'pedagogical' processes of socialisation and enculturation. The difficulty is not with holding a psychological discussion but with the assumption that psychology is a science of individuals. This 'individualist' presumption results in the charge of reductionism being leveled at social theories that discuss or draw on psychology. The criticism is that such social theories place to much weight on individuals and their psychology at the expense of social and cultural context. However, in fact, there is little reason to assume psychology is necessarily individualist in any reductive sense and many 'psychological' perspectives recognize the importance of the social dimension, particularly with regard to learning, development, socio-cultural practices and, increasingly, the nature of the mind.

Nevertheless anthropological and sociological theorists tend to be reluctant to discuss any explicitly psychological aspect of their accounts. Situated within both anthropology and sociology Bourdieu's perspective has been understood in this way. Bloch suggests that, with the habitus, Bourdieu "reinvents psychology [as] a kind of private psychology for the use of social scientists" (2005, p. 19). However, in a relatively late work, Bourdieu is clear that "sociology and psychology should combine their efforts … [and] overcome their mutual suspicion" (2000, p. 166). Cognitive anthropology is one attempt to effect a reconciliation between the social, or socio-cultural, and the psychological. This chapter draws on research in this domain, particularly the reconceptualisation of apprenticeship.

In the first part of this Chap. 1 introduce the notion of 'practice' and construe thinking as a practice. I then trace some of the implications of a fully social or cultural psychology with regard to medical ethics education and professional reproduction. I recall some of the terms set out in Chap. 1 and the Bourdieuan social theory discussed in Chap. 2. Following a brief discussion of Vygotskian psychology that underpins all of the perspectives discussed in this chapter I turn to

N. Emmerich, *Medical Ethics Education: An Interdisciplinary and Social Theoretical Perspective*, SpringerBriefs in Ethics, DOI: 10.1007/978-3-319-00485-3_5, © The Author(s) 2013

the renewed theorization of apprenticeship offered by socio-cultural psychology and cognitive anthropology and its perspective on the 'situated' nature of cognition and learning. Finally I reexamine the notion of thinking dispositions and their (re)production in the light of my discussion.

5.2 Thinking as Practice: A Socio-Cultural Perspective on Medical Ethics Education

As a further example of the practice turn in social theory (Schatzki et al. 2000). Talja (2010) notes that it is central to Lave's socio-cultural perspective but that it is a relatively neglected by those who have taken up the 'Communities of Practice' perspective (Lave and Wenger 1991). Their focus has been on 'the community' rather than on the variable range of practices that define the community. Schatzki suggests that underling the practice turn is an attempt to move "beyond current problematic dualisms and ways of thinking... [and to contend] that practices at once underlie subjects and objects, highlight non-propositional knowledge and illuminate the conditions of intelligibility" (2000, p. 11). Practice perspectives are an attempt to move past epistemologies that are restricted to or privilege propositional knowledge. They attempt to move past a focus on what Ryle called 'knowing-that' to include tacit knowledge, or 'knowing-how.' This is the case not only in Bourdieuan social theory (Gerrans 2005; Taylor 1993) but also across a range of theorists that have made variable use of the term across a variety of domains including Brinkmann in psychology (2010), MacIntyre in moral theory (1981) and, in perhaps its broadest application, by Taylor in his philosophical reflections on social theory (1985a).

Whilst there is no single account of 'practice' the idea of tacit knowledge, of embodied knowing-how, is consistently accorded centrality in understanding the array social and cultural activities of human beings, including those activities for which propositional knowledge appears essential. If we move away from a purely corporeal sense of embodiment we can begin to consider how the conscious use and manipulation of propositional knowledge is reliant on an array of non-conscious and tacit knowing. Roth suggests a good criteria for embodiment in this sense is "more or less of what a person knows [which] is either inaccessible in principle or not thematised in everyday situations because the matter 'goes without saying'" (Roth 2001, p. 471). Such knowing is, of course precisely what Bourdieu suggests is embodied in the habitus and calls 'a feel for the game.' Arguable such knowing also underpins everyday ethical action (Dreyfus and Dreyfus 1990, 1991, 2004). Lave's own 'practice turn' also reflects a concern for tacit knowledge and she expresses surprise at the way in which learning has been narrowly focused on propositional knowledge suggesting that:

> "[w]hat seemed far more startling [than the assumption that theories of learning ought to be solely about psychological processes] is the incredibly narrow pervasive history of

philosophical and later psychological treatments of 'learning' as wholly an epistemological problem—it was all about knowing, acquiring knowledge, beliefs, skills, changing the mind, moving from institutions to rules, or the reverse, and that was all" (Lave 1996, p. 156).

In the context of education a focus on practice facilitates recognition that learning to use formal or propositional knowledge is necessarily accompanied by a degree of tacit knowledge; learning is not, simply, the passive acquisition of knowledge but involves the active participation of the student in classroom exercises that make use of the knowledge. Learning to know-that X is not simply to acquire a fact, a proposition, a belief or a rule but, rather, it is to acquire the know-how that attends its use and, furthermore, that attends its use in a particular context. Lave is expressing surprise at the fact that theories of learning do not appear overly concerned with the way in which this know-how is learnt and how this relates to classroom teaching. In terms of our previous discussion of medical ethics education we might suggest that the hidden curriculum of medical education is not only found within the clinical context, the more informal 'apprenticeship' aspects of medical education, but also in the classroom. In the first instance, the tacit lessons that might be learnt in the formal medical ethics classroom may include: the relationship between clinicians and medical outsiders, particularly philosophers or 'ethicists;' the scope of abstract ethical discussion as compared to its practical implications in the clinic; the nature of ethical decisions and when medical doctors need to ensure the collective agreement of a clinical team; and the proper way to frame and discuss the ethical dimensions of a case. All these can be considered examples of the informal or hidden curriculum.

However socio-cultural perspectives also insist that learning is situated and that the situation in which learning takes place can be an essential component of the practice being taught. As a result they insist that something learnt in one context cannot be simply transferred to another context; there must be some degree of 'relearning.' This is not a simply process of (re)applying knowledge but, rather, one of re-appropriating and resituating the cognitive practices of skilful minds (Gellatly 1986). Therefore, in the second instance, the tacit lessons being taught in the formal medical ethics classroom concern the know-how that is involved in making use of the propositional knowledge of medical ethical concepts and principles to make ethical judgments and (re)form ethical beliefs. Following a rule is not a simple matter but involves a form of knowing-how that dictates the way in which a rule is to be applied and understood in a particular context, culture or community of practice (Taylor 1993; Stueber 2005). Medical education involves an implicit instruction on the nature of ethical principles, their applicability and scope. Some of this can be made explicit but much of it remains tacit. Through participating in a dialogical use of the principles of medical ethics the medical student is initiated into the practices of ethical rule following: the application and use of ethical rules that constitutes modern medical ethics.

Much of the educational research that has been done into the tacit knowledge attendant on learning to make use of propositional knowledge has focused on mathematics. Arithmetic is seen as a tool for investigating relations between culture and cognition (Reed and Lave 1979) and ideal ground for researching the situated nature of everyday cognitive practice(s). Arithmetic can be rendered in a context-free manner and is thereby clearly specifiable for methodological purposes. With this in mind researchers have explored the way in which people use mathematics when accomplishing everyday, routine and mundane calculations (Lave 1988). Furthermore mathematics education is a central aspect of research into the nature of formal learning (Atweh et al. 2001; Sfard 2007, 2010; Sriraman and English 2010). Such research raises questions about the characterization of cognition as context-free (Lave 1988, p. 92).

An example of the insights it offered by this research can be seen in an episode of The Wire where a child is shown attempting to solve an arithmetic problem. The problem he is concerned with is of a standard type and involves counting the number of people who get on and off a bus at a number of stops. The child is unable to calculate the correct answer. An older boy rephrases the problem in terms of drug deals where upon the child immediately provides the correct answer. Perhaps to the pleasure of educational traditionalists the reason the child gives for his success is that were he to get 'the count' wrong whilst dealing drugs he would be beaten for it. However from the perspective of situated learning the child has learnt to perform mathematical calculations in a particular context and for a particular purpose—dealing drugs—but he is unable to transfer those same 'cognitive functions' to the arena of bus passengers or, indeed, to abstract mathematical calculations more generally. Situated learning theory challenges the assumption that cognitive skills are easily transferable. Whilst it does not deny outright that cognitive skills might be transferable it insists that, in being 'transferred', such skills must become 'reacclimatised' through their exercise in a different situation. The transfer of cognitive skills from one domain to another is not a simple matter and we should appreciate that, in being reacclimatised, they—or the uses to which they can be put—are quite literally *relearnt*.

The idea that learning something in the abstract is insufficient guarantee that it will be transferred into practice is widespread in medical ethics education. As Guile and Young's claim: "formal learning can be enhanced if the skills and knowledge that students learn are embedded in a[nother] social and functional context" (1998, p. 179). A commitment to this perspective can already be discerned in medical ethics education. As the lessons learnt by medical students in a classroom situation have to be put into practice in another situation—the clinic—then we might expect there to be a case for the extension of medical ethics education into the clinical context and, indeed, this is what we find (Miles 1989; Bicknell 1985; Goldie 2000). There is an imperative in medical ethics education to go beyond the classroom and to become integrated into the clinical context as well as to make use of real life cases rather than the artificially constructed 'thought experiments' preferred by philosophy and applied ethicists.

However the practice of medical ethics is less specified, and specifiable, than those who have used the practice of arithmetic to challenge to primacy of abstract learning situated in the classroom for the conduct of concrete calculations in the supermarket (Lave 1988, Chap. 7). Furthermore, as an aspect of professional practice medical ethics is not only located within clinical practice, in the dynamic context of reflection-in-action, but also within the more contemplative moments of reflection-on-action. At its best, medical ethics is both a contemplative and a practical activity or, in Aristotelian terms, both theoria and praxis. As such we might suggest that medical ethics is a theory that is not only put into practice but is, itself, practiced. It is important to recognize that a theoretical activity is still an activity and one which is underpinned by specific cultural and cognitive practices. This is particularly true when, like medical ethics, the theoretical activity is meant to have practical consequences.

This perspective is in accordance with the perspective that the "activity, and its values and goals are generated simultaneously" (Talja 2010, p. 207). Furthermore Lave thinks similarly for 'problems' and their 'solutions.' Problems are not merely identified with a solution subsequently attempted, rather the identification and definition of a problem is always accomplished for the purposes of attempting a particular solution. Lave has identified a tendency to equate cognition with 'problem solving.' It is easy to assume that the cases of medical ethics are, in a sense, problems to be solved. Furthermore, the structure of medical ethical cases echoes that of the patient, the 'medical case' more generally (Anspach 1988). Medically speaking patients are constituted by their case histories, their symptoms and any relevant social factors that, taken together, are used as the basis of a diagnosis which must then be treated and, thereby, 'solved.'[1] The tasks that face medical professionals in regards the ethical dilemma and the patient is not so much the provision of a solution but, rather, diagnosing/defining the problem by making use of the tools—the purported solutions—they have to hand, medical treatments/the concepts of medical ethics. In both medical ethics education and medical education more generally the task is to equip medical students with the appropriate tools with which they can structure the very problems that appear to independently precede them.

The implication of the practice perspective is that thinking, a form of cognition, is best understood as an activity and that concepts can be understood as tools that the mind makes use of in undertaking this activity. What this reveals is that in applied ethics, philosophy and perhaps academia itself we often mistake the products of our thinking for activity of thinking itself. The practices of academic applied ethicists produce journal articles: extended discussion and arguments regarding the rights and wrongs of a particular topic, question or act. However no one simply produces such linear monological arguments in a *de nova* fashion. Rather they have been gradually constructed, developed and refined through a process of, the *activity* of, dialogical reason and reflection. Whilst medical students are inducted into an activity similar to that of the applied ethicist the ends of those activities are markedly different. The ethicist aim is the articulation and defense of

---

[1] Although I am reminded of Irwin's distinction between General Practice and hospital medicine this is a somewhat misleading perspective. Encouraging medical students to see patients as, simply, problems to be solved is approximately as helpful as seeing the ethical dilemma in the same way.

an argument that supports or critiques an ethical position—other ethical arguments—in a manner that meets the substantive criteria of the discipline. In contrast the medical professional's aim is, primarily, the facilitation of (ethical) medical practice. We could compare this difference to the idea of a standard of proof. In criminal cases the standard of proof required is 'beyond all reasonable doubt' whilst in civil cases it is 'the balance of probabilities.' Neither is easily or clearly defined but we do have a sense of the difference and its meaning. Expanding this metaphor we might conclude that the ethical practices of medical professionals are directed at producing an ethical perspective that, on the balance of probabilities, indicates the right thing to do whilst the practice of the applied ethicists is the articulation of a robust defense or critique an ethical position that attempts to place the conclusion beyond all reasonable doubt.

However this is to maintain a focus on the formation of substantive positions and making specific judgements. Whilst this does occur in everyday medical practice substantive judgements are also embedded in the ethical structures of the medical profession; medical doctors do not have to decide if they can actively euthanise a patient because such actions are ruled out by the professional guidelines, guidelines they encounter during medical school. Furthermore medical professionals do not confront clearly defined ethical issues as if for the first time every time they confront them. For example virtually all medical professionals, particularly GPs, will have a reasonably well thought out position on the ethics of abortion. Furthermore those who personally object to the abortion will also be aware of their responsibilities to refer patients who are seeking abortions to one of their colleagues. There will be little need for them to rehearse the arguments when a patient comes to them seeking an abortion. In everyday medical practice making judgements is not the primary use of the tools of medical ethics. Rather healthcare professionals use them to structure the moral landscape of their practice in a manner that accords with the ethical guidance issued by the profession as a whole. Whilst it is the case that further reflection and analysis of particularly troubling cases is sometimes required on occasion this is one of the primary uses of ethics in everyday medical practice.

The ethical education of medical students introduces the ethical structures of the profession and encourages them to form their own substantive ethical beliefs such that they can, explicitly or implicitly, rely on them during future medical practice (although, of course, they remains open to on-going revision and refinement).

This is an important aspect of medical ethics education and it has more in common with the practices of applied ethicists that does the ethical practices found in everyday medicine. What is interesting, however, is how this more formal educational exercise results in the ethical enculturation of medical students. Medical students can rely on the judgements and beliefs formed at this time and the ethical structures of the medical profession. This indicates the dispositional nature of such beliefs. Medical students also begin to develop a sense of when further reflection might be required. The process of medical education produces within students what, in the context of developing Bourdieuan social theory,

Brubaker has called an "unreflective disposition to reflect" (Brubaker 1993, p. 225). This broad disposition underlies the more specific idea of thinking dispositions discussed in Chap. 2 and to which I return to in the final section of this chapter. First, however, I turn to a discussion of Vygotskian psychology and the rethinking of apprenticeship that has attended the development of socio-cultural theories of learning.

5.3 A Vygotskian Perspective

In common with Bourdieu's approach, socio-cultural theories of learning are focused the idea of practice. However, unlike Bourdieuan social theory, they also directly draw on psychology for inspiration in articulating their perspective. In extremely broad terms psychology can be roughly divided into two camps or traditions: Piagetian and Vygotskian. Whilst Piaget and Vygotsky were contemporaries, and did in fact correspond and meet, the cold war and Vygotsky's early death conspired to delay the translation and impact of his work. Thus in non-Russian speaking countries Piaget has been, and remains, the formative and dominant influence whilst Vygotsky, even now, remains somewhat neglected (Roth and Lee 2007). Kohlberg's (1981) moral psychology has, until recently, been the dominant and pre-eminent paradigm in moral psychology and his research program is firmly located within the Piagetian approach drawing directly on the groundwork he laid (Piaget 1997). Furthermore, challenges to Kohlbergian moral psychology have rarely drawn on the Vygotskian perspective for inspiration. For example Carol Gilligan's (1993) influential critique is focused on the (gendered) nature of morality and its definition or 'operationalisation' in Kohlbergian research and not on challenging the underlying social-psychological assumptions.

The Piagetian perspective tends to dichotomise the individual and their social context. Whilst psychological development is seen as contingent on having a social context it is also presented as a universal quasi-biological phenomenon that is merely facilitated by social or cultural influences and proceeds in a linear fashion. This perspective can clearly be seen in Lawrence Kohlberg's stage theory of psychological moral development. Famously Kohlberg suggested that his theory would be disproved if any individual were found to have skipped one of his stages of development or traversed them in an order different to the one he proposed. Given the nature of this prescription it is difficult to imagine that, given its variability, culture or society can be accorded any great influence in this view of psychological moral development.[2]

---

[2] The reader should recall that the account of development in play in this work is that of the participationists (see Sect. 1.2: Acquisition and Participation. In this view development does not suggest the internal transformation of people as much as it suggests a change in their practices, their forms of doing. Thus, contra the standard presumptions of developmental psychology, it proceeds across the life course.

Furthermore the Kohlberg/Piaget approach is focused on moral judgement. Abend (2012) has suggested the approach adopted by moral psychology is 'judgement-centric' and neglects an array of morally relevant activities. Psychological research that adopts a MJA excludes a wide range of our moral thinking, action and behaviour from its purview and therefore gives a limited account of our moral psychological being. Abend argues that there is a need to adopt a variety of perspectives and approaches to researching our moral and ethical lives if we are to get a proper grasp on the object(s) of our concern. He further suggests developing a wider appreciation for the sociological in the psychology and this is something that can be furthered by adopting an explicitly Vygotskian approach. In a widely cited passage Vygosky says that:

> "Any function in the child's [our] cultural development appears twice, or on two planes. First it appears on the social plane, and then on the psychological plane. First it appears between people as an inter-psychological category, and then within the child [ourselves] as an intra-psychological category" (Vygotsky 1981, p. 163).

Whilst neither Vygotskian psychology nor I offer a panacea for the problems identified by Abend it can assist in broadening the focus beyond the making of moral judgements and the associated developmental psychology. The importance of recognising the influence of the cultural plane is reinforced by my focus on medical ethics education. Furthermore my discussion concerns the education of medical students who, lest we forget, are adults and not children. Of course as young adults we might suppose that medical students are still settling into their role as mature members of the moral community nevertheless there is little need to treat them as if they were still undergoing some form of basic moral development or, therefore, to think of their medical moral development in the same or similar terms as many seem to have done (Chalmers et al. 2011; Hren et al. 2006, 2011; Lin et al. 2012; Lind 2000; Patenaude et al. 2003; Self et al. 1988). Rather, as the ideas of professional socialisation and, particularly, adult enculturation suggest, the moral and ethical development of (young) adults is a process of (re)acclimatising to a new moral and ethical context and culture. Formal ethics education contributes to this process through, in part, presenting medical students with a new conceptual language of ethics and its associated reasoning patterns. This is not simply a mater of individual (psychological) development. Such learning cannot be construed as a purely individualised process as it is not only a matter of conceptual acquisition but also of collectively participating in ethical reasoning. Medical students acquire the concepts and reasoning skills of modern medical ethics through participating with each other in their use.

Grenfell questions "whether the intra and inter-psychological are not simply habitus and field by another name" (2009, p. 443) and whilst I would disagree— they are not *simply* the same—the comparison is instructive. Strauss and Quinn present their understanding of Bourdieu in not dissimilar terms, suggesting that "habitus, (intra-personal knowledge) is structured by objects and practices in cultural learner's public environments (the extra-personal realm)" (Strauss and Quinn 1998, p. 45) and this view is more satisfying if we recall that our intersubjective awareness of others who are also considered to be intersubjectively aware is fundamentally constitutive of the extra-personal realm but also the intra-personal realm. We come to recognise ourselves as we come to recognise others. In the Vygotskian view we should then appreciate that reflection and metacognition as an aspect of our cultural development that must first appear *between* individuals, usually between the initiated and the uninitiated, before it can appear *within* them. Furthermore, given the discussion in Chap. 1, we can appreciate that metacognition is not an abstract cognitive ability that is given and subsequently exercised in relation to new theoretical perspectives and knowledge. Rather, working in the light of previously developed metacognitive abilities, this generically named skill can be relearnt and reacclimatised to new contexts and to new 'ways of knowing' that it seemingly precedes.

At this juncture we might consider Vygotsky's 'zone of proximal development'[3] defined as "the distance between the actual developmental level as determined by independent problem solving and the level of potential development as determined through problem solving under adult guidance or in collaboration with more capable peers" (Vygotsky 1978, p. 86). In the context of medical ethics education these more capable peers might include other students,[4] medical educators, healthcare professionals, philosophers, theologians, sociologists, ethicists, other professionals involved in the ethics education of medical students and, where included, patients and those with other forms of relevant expertise e.g. Disability activists. The 'zone of proximal development' is an area of developmental 'space' that, given the individuals intra-psychological functions, can become occupied though inter-psychological assistance, engagement and 'scaffolding' (Wass et al. 2011). The development of some cognitive abilities is predicated on previously acquired cognitive abilities; no medical student could learn to think modern

---

[3] This is the most widely used translation but it has also been called the zone of 'next' or 'potential' development. The zone of next development is used relatively rarely as it indicates a linear and overly prescribed notion of development more consistent with a Piagetian perspective. In contrast a potential development may never occur whilst a proximal space may not ever become occupied, indeed other changes (developments) might come about that remove a previously potential psychological development from proximity.

[4] It is worth, briefly, noting what it means to consider students each other's 'more capable' peers in this context. It is not simply that a more advanced or competent student can lead other students to have new insights or make new connections that they have already achieved. Students are taught together because together they can develop in ways that they could not achieve on their own. The case at hand, medical ethics, collective discussion amongst peers can lead to an increased development, an increased competence, in medical ethical reflection on the part of individuals. Groups can dialogically bootstrap the (monological) individuals that constituted them. Furthermore, individuals can attempt to go beyond their own monological perspective by engaging others either directly or indirectly, through the media of materials such as text-books, published articles or any other forum for ethical discussion. As adults and experienced or practiced learners medical students can, to a degree, guide and direct their own development. I think this is particularly true in the case of moral and ethical development that attends medical education as, after all, they are mature individuals indicating that, in our broader perspective on moral functioning, engage with, and not simply judge, the moral situations that confront them.

medical ethics without first having undergone the process of moral development that occurs during (western) childhood. Ethical enculturation, of whatever kind, cannot precede our basic moral socialisation as it provides the basis for our basic moral perception on which the more formal task of ethical reflection is predicated. However, as implied by Zigon's (2008) theory of the relationship between ethics and moral dispositions, it is often the case that the more basic ability is altered and further developed by the subsequent activity that is predicated upon it.

Mercer suggests we see "language as a psychological tool for organising our individual thoughts [and that in] childhood a fusion of language and thinking occurs which shapes the rest of our mental development" (2000, p. 10). Furthermore linguistically formed concepts are not simply the objects of thought but, rather, the "structural elements of thoughts" (Gillett 1987, p. 101). We think with and within the conceptual structure of language. As discussed in Chap. 1, meta-cognition is a conceptual, linguistic and propositional form of cognition that first appears as a social and dialogical phenomenon prior to its appearance as an individual (and monological) phenomena. The purpose of formal medical ethics education is to enable medical students to engage in this particular form of ethical reasoning. They way in which this is accomplished is through encouraging medical students to socially and dialogically engage with medical ethics and to participate in this form of reasoning. Cole (1985) considers the zone of proximal development to be the psychological space in which culture and cognition create one another thereby generalising the concept for all human development and not just childhood development (Guile and Young 1998, p. 179). The cultural practice of modern medical ethics is brought within the medical students zone of proximal development through their attempts to (re)cognise this practice and engage in it themselves. Ethical enculturation results from bringing together the ethical culture of medicine and the existing cognitive abilities of medical students within an educational environment structured by more capable peers (teachers, educators, role models etc.) and, simply, equally capable peers i.e. each other. Whilst formal medical ethics is not appropriately applied to everyday life we might expect medical students formal ethical enculturation to impact on their moral perception and ethical deliberations beyond the clinical context. What this reveals is that development is a complex and reflexive process embedded within a particular socio-cultural context.

This view can be considered a riposte to those who, in a Kohlbergian frame, have raised concerns regarding the apparent plateauing, if not outright 'regression', of medical students moral reasoning or, more accurately, ethical judgement (Brockett et al. 1997; Hren et al. 2011; Lind 2000; Patenaude et al. 2003). Without mounting a direct attack on the Piagetian ideas of childhood (moral) development per se we can resituate our understanding of medical student's moral and ethical development outside of this frame. The ways in which medical students are taught to talk and think about the ethics of medical practice is part of their development into professionals. They are being introduced to a new practice, a new way of doing and thinking ethics that is appropriate to the culture of medicine. As such it is likely that medical students will, as part of their developing professional ethical abilities, initially rely on the more orthodox or 'conventional' positions and approaches based on the authority and social structure of medicine. Only having fully comprehended these positions and the associated reasoning patterns will they have sufficient familiarity with the moral terrain of medical practice and medical ethical reasoning to independently undertake the activity of ethical reflection such that they might come to their own conclusions.

Of course this is not to say that medical students will in fact subsequently develop their own independent, possibly heterodox, ethical perspectives predicated on their own individual and professional moral and ethical authority. Many may simply rely on the established ethical structures of modern medicine. However, if they are to do so, medical students must first develop their own (sense of) professional authority and the stating points for this development are the authoritative perspectives that characterise i.e. socially structure, the medical profession. Such perspectives include not only concrete ethical judgements about, say, active euthanasia but also include the second order, or meta-ethical, forms of normativity such as the norm of medical ethical reasoning done in a certain way, with certain tools or concepts. Coming to recognise the authoritative ethical positions of the medical profession and, perhaps more importantly, the normative structure that medicine expects ethical justifications to take necessarily precedes the production of fully independent professional ethical reasoners. Furthermore, in practice, obedience to these authoritative positions and the normative structure of ethical reflection, judgement and justification is a condition of professional membership and the maintenance of that membership. Of course the ethical structures (guidelines, rules, and so forth) of the profession do not and cannot fully determine all of the moral questions and concerns raised by medical practice. Thus, in practice, there is a requirement for medical professionals to be capable of coming to independent or 'post-conventional' moral judgements and medical ethics education rightly aims at producing such individuals. Nevertheless we must remember that, as Bourdieu suggests, "the only form of durable freedom [is] that given by the mastery of an art... which is acquired, paradoxically, by obligated or elective submission" (1999a, p. 340).[5]

What, and how, we know and reason is not independent of society, culture or our environments but is given to us to use both through and within those environments. In the first instance this concerns our learnt understanding, i.e. how medical students learn to reason with the ethical concepts they are given. In the second instance this affects our ongoing abilities even after we have become competent and even expert practitioners of a particular cognitive task. Following Bourdieu we should "not contrast practice with learned understanding; ... [but realise that] we cannot understand practice without rethinking the nature of learned understanding" (Strauss and Quinn 1998, p. 45). Such a rethinking is offered by the socio-cultural perspective of Situated Learning Theory (SLT) which has

---

[5] This paradox, the attempt to produce individuals whose moral perspective is emancipated from those who have instructed them, is found throughout moral education (Peters 1963).

emerged from a series of related research programs in social and cultural psychology, cognitive anthropology and apprenticeship theory, all of which are indebted to a Vygotskian psychological perspective.

5.4 Re-Thinking Apprenticeship: Rethinking Medical Ethics Education

In their book 'Situated Learning' Lave and Wenger suggest that, circa 1988, there was a desire to "rescue the idea of apprenticeship" (1991, p. 29) particularly at the Institute for Research on Learning at Paulo Alto. The aim was not, simply, to rethink the idea of an apprenticeship as a social institution but as the basis of a social theory of learning (Guile and Young 1998). Their own view recasts apprenticeship as the legitimate peripheral participation of novices in a community of practice and is an attempt to overcome "the assumption that learning is an individual process, that it has a [clearly delineated] beginning and an end, that it is best separated from the rest of our activities, and that it is the result of teaching [alone]" (Wenger 1999, p. 3). Furthermore the investigation of apprenticeship learning led "to the conclusion that the 'informal' practices through which learning occurs in apprenticeships are so powerful and robust that this raises questions about the efficacy of standard 'formal' educational practices in schools rather than the other way around" (Lave 1996, p. 150). Thus the rethinking of apprenticeship sought to not only reconsider the presumed disunity of formal and informal learning process but also to suggest that informal learning processes attended formal pedagogic endeavours. The pedagogic practices of formal education become understood as a particular cultural context for learning and that all learning, including classroom learning, was 'situated;' no context in which learning takes place can be considered pedagogically or epistemologically neutral. Furthermore one purpose of renewing the idea of apprenticeship was so that the insights it generated could be reflexively applied to the classroom. This fact, that formal education and informal learning are not dichotomous or fully distinct, implies that neither can be construed as intrinsically 'better' or 'worse' than the other and indicates that this research program has something of a political edge. Strauss (1984) argues that to overcome the dichotomy is to overcome an ethnocentric assumption in the anthropology of knowledge and cross-cultural psychology.

Whilst the knowledge being taught in classrooms might be decontextualised, generalised and abstract the decontextualisation, generalisation and abstraction is itself understood as a particular cultural form. Similar to the Vygotskian understanding of psychological development the situated learning/apprenticeship perspective is thoroughly socialised (de-individualised), contextualised (seen in relation to the rest of our activities) and open-ended. It is then unsurprising to find that just as the concept of psychological development has been stretched into adulthood rethinking the concept of apprenticeship has entailed applying it to childhood. In her 'Apprenticeship in Thinking' Rogoff (1992) demonstrates how the individual and the social world are mutually constitutive, how guided partic-ipation (an activity in which children play an active role) marks many activities in which learning takes place, and how shared problem solving and collaborative thinking is the basis for cognitive development.

Lave and Wenger admit to having concerns with the conceptual stretching of apprenticeship and, again indicating the way in which the rethinking of appren-ticeship is bound up with rethinking learning per se, suggests it may have become a synonym for 'situated learning' itself (1991, p. 30–31). These concepts have much in common as they are involved in the development of the idea that learning is not simply a matter of transmission but a socio-culturally situated processes involving participation contributed to the reexamination of apprenticeship learn-ing. As a counter to Lave's concerns about the synonymous nature of appren-ticeship with situated learning, and therefore the potential redundancy of the former, we might note that Rogoff considers her use of the term 'apprenticeship' to be metaphorical. She suggests using the term directs attention to neglected aspects of childhood development including the active role of the child as well as the relevance of culture, society and role of interlocutors, other people who engage with the child (1992, p. 192). We might suggest that this metaphorical usage extends to the idea of a cognitive apprenticeship.

It is, I think, useful to recognise the metaphorical structure of the broad idea of a 'cognitive apprenticeship' described by Brown et al. as a methodical attempt "to enculturate students into authentic practices through activity and social interaction in a way similar to that evident—and evidently successful—in craft apprentice-ship" (Brown et al. 1989, p. 37). However in the case of medical (ethics) edu-cation, and in the context of forging a connection between informal and formal learning, we can appreciate that medical students are in fact apprenticed to medicine and this includes an explicitly cognitive aspect; they are being trained to think in a certain way(s). Nevertheless, to reflexively anticipate and apply some of the arguments presented below, we must recognise that in its application to the cognitive domain the concept of an apprenticeship is a tool for thinking with.

One direction in which this thinking tool can be taken is with regard to the two concepts of medical ethics discussed in Chap. 3. In that chapter I argued that, within everyday medical practice, something of the critical function of medical ethics had become absorbed into traditional medical morality. In this view the concepts and reasoning associated with modern medical ethics are a mundane part of practice, a language used to talk about and frame ethical problems and, where appropriate, to articulate and debate substantive positions. In the arena of everyday medical practice the relationship between medical ethics and medical morality has ceased to be one of critique and has (or is) becoming the traditional mode of ethical perception and articulation. My theoretical concerns are, then, not focused on any supposed radicalism of medical ethics education but, rather, with its standardisation. Conceiving medical ethics education as a cognitive apprenticeship is to accept that its aim is to produce medical professionals who approach the ethical issues of medicine in the same or similar manners. Doing so does not fully determine the substantive positions that may be adopted in relation to a specific issue, however, it does determine how such ethical issues will—or can—be framed and discussed in the professional context.

As I have suggested the cognitive apprenticeship of medical ethics is only one aspect of the medical apprenticeship and the ways in which medical students are being taught to 'think' in the professional context. In Chap. 1 I briefly discussed the influential notion of reflection and the way in which it has been used to understand professional practice and education. The cognitive apprenticeship of medical ethics education is an example of a reflective or, more accurately, a metacognitive approach to professional education. This indicates the importance of concrete medical ethical cases to formal medical ethical education. Medical students need realistic examples which they can examine using the ethical concepts of modern medicine. Through this dialogical exercise they develop their ethical metacognitive abilities that are then taken back into the clinical context and further developed as appropriate. Using their conception of apprenticeship (and situated learning) as involving 'legitimate peripheral participation' in a 'community of practice' Lave and Wenger suggest that "the purpose is not to learn from talk as a substitute for legitimate peripheral participation; it is to learn to talk as a key to legitimate peripheral participation" (Lave and Wenger 1991, p. 109). One could take this as an indication that the purpose of formal medical ethics education is simply to provide medical students with the concepts of medical ethics and that the 'true' learning or apprenticeship does not occur until they are used in the clinical context. However this would be to, first, underestimate the relevance of 'legitimate peripheral participation' and situated learning to formal, classroom based education and, second, to fail to recognise that the more abstract metacognitive practice of the formal medical ethics classroom are essential to, and increasingly replicated within, actual clinical practice. A medical ethics classroom can be, as discussed, considered a Community of Inquiry (Lipman 2003; Mercer 2000) in which medical students participate and so become inducted into a way of thinking.

Formal medical ethics is a practice primarily characterised by metacognition and, therefore, by reflection-on, rather than in, -action. Certainly it is important that ethical awareness is maintained in action but, nevertheless, formal medical ethical reflection is a practice best accomplished away from actual clinical practice. The Clinical Ethics Committee (CEC) provides an instructive example as well as highlighting the fact that, rather than being culturally neutral, ethical reflection-on-action is essentially characterised by medical culture.

The CEC is constituted to provide clinicians with a forum for the discussion of ethical issues. When consulting a CEC clinicians are not seeking to be told the ethical solution to their problem rather they are seeking to engage in ethical debate, discussion and dialogue. In order to be motivated to consult a CEC the clinician must have identified an ethical problem to the extent that they can express it to others. In itself this will have involved exercising their ethical faculties that are further deployed in the context of the CEC. We might suggest that, culturally, the CEC exists somewhere between the 'informality' of clinical practice and the 'formality' of the medical ethics classroom, the formality of an ethical code or set of guidelines, or, indeed, the 'formality' of textbook or journal article on the subject. We can see here that formality/informality is being used as indicating the degree to which 'culture' is present and not whether or not culture is present. In educational contexts it is this dichotomy that Lave, Wenger, Rogoff, Guile, Young and others have attempted to overcome by rethinking the concept of apprenticeship and the process of learning. What it reveals is that the abstraction and generalisation inherent in modern medical ethics should not be taken as indicating that they are culturally neutral concepts. The 'objectivity' of medical ethics is predicated on professional distance, the concept of the patient and their relationship to doctors. In practice, in theory and in pedagogical delivery modern medical ethics is a socio-culturally situated phenomenon.

In Chap. 2 I discussed the idea of thinking dispositions and how they were required to develop a cognitive account of habitus. In this section I have highlighted the way in which the concept of an apprenticeship can be used to understand how particular cognitive practice are learnt or enculturated. This perspective also involves an understanding of thinking that sees it as an activity and concepts as tools which our minds make use of. As with any other tools we can make use of them in better or worse ways depending on our level of skill, our know-how, the tacit understanding without which the practice could not be accomplished. This understanding is built up through repeated use of the tools by novices and, in particular, through a legitimate peripheral participation in the expert use of the tools by experts. This idea is more intuitively obvious in the case of physical rather than cognitive practices. We can easily comprehend how trainee carpenters become increasing adept at using the tools of their trade as their level of their (guided) experience increases. Given the rethinking of apprenticeship this can no longer be consider simply the result of socialisation but also includes enculturation. Whilst in the context of traditional apprenticeships, an experience that is predominantly structured by informal learning, it is the case that the latter is more prevalent than the former. However the opposite is true in regards a cognitive apprenticeship, an experience that is predominantly structured by formal learning. As Brown et al. suggest:

> "To learn to use tools as practitioners use them, a student, like an apprentice, must enter that community and its culture. Thus, in a significant way, learning is, we believe, a process of enculturation" (1989, p. 33).

In considering medical education as an apprenticeship with a cognitive dimension we can recognise the importance of both socialisation and enculturation, both formal and informal learning. Overtime these processes and their effects become inseparably intertwined something we can perceive with respect to the moral socialisation and ethical enculturation of medical students. Whilst I do not wish to suggest that traditional medical morality and critical medical ethics never conflict the embedding of the latter within the former has brought them into a more complimentary relationship than was previous the case; the theoria of contemporary professional medical ethics is, for the most part, in line with the practices of modern medicine. Medicine and medical professionals remain, of course, capable of committing ethical errors however, broadly conceived, the moral landscape of modern medicine is in line with its ethical structures. If we considering professional practice and medical education in these broad terms then we can appreciate that they are long-term projects that involve sustained and on-going activity. This further reveals the way in which theory and practice are intertwined, particular in the case of medical students education. Tishman et al. suggest that teaching for thinking dispositions:

> "combine[s] theory and practice into one rhythm of effort. In practical terms, that means taking either a think-do-think approach to enculturating thinking dispositions in the classroom, or a do-think-do approach. Both of these approaches involve cycles of reflection and practice" (Tishman et al. 1993, p. 152).

Medical students do not, of course, officially 'do' anything but they do legitimately and peripherally participate in medical practice. Their formal and informal learning is layered, each builds on the other over time. Moral socialisation and ethical enculturation is similarly layered and medical education is, to recall Bourdieu's phrase, an extended collective enterprise of inculcation. The apprenticeship view not only suggests both processes, socialisation and enculturation, are active in this enterprise but also suggests that they interrelated. However, in the final section of this chapter, I first discuss what, precisely, is enculturated as part of the formal ethics education of medical students.

5.5 What is Enculturated by Medical Ethics Education Considered as a Cognitive Apprenticeship?

The theoretical construction I have offered suggests medical ethics education can be considered as cognitive apprenticeship and that the pedagogical success of this apprenticeship is achieved through a process of enculturation. This perspective is, obviously, to be situated alongside broader understanding of medical education that rely on the more common notion of socialisation (Luke 2003; Sinclair 1997; Brosnan 2008). On this basis it is worth directly addressing what we can consider to be enculturated by formal medical ethics education. There are a number of distinguishable objects we might consider including: concepts; cognitive skills; beliefs; and thinking dispositions. I consider the first three to be relatively straightforward cases of enculturation. However the matter is more complex in relation to thinking dispositions as it is uncertain what, if any, ontological commitments they might imply (Splitter 2010).

5.5.1 Enculturating Concepts

Given the philosophy of language we have been operating with the fact that 'words' or 'concepts', particularly those more 'abstract' or 'theoretical' concepts which have a role in ethical metacognition, are enculturated is a reasonably non-controversial statement. If meaning is given through use, and participation in the use of theoretical concepts requires (or at least implies) (meta)cognitive activity, then it is certain that the acquisition of such concepts involves an active process of inquiry, something which is indicative of enculturation (Shimahara 1970, p. 148). Within the view I have articulated, medical ethical concepts are clearly cultural; they are tools for 'doing' or practicing medical ethics within modern medical culture. Being able to use and think with them in a culturally appropriate manner is a signal that one belongs to the culture. Therefore engaging in reflection in order to debate the ethical aspects of particular cases and dilemmas, something which includes the formation of ethical judgements, is an "enculturative experience [as it] involves the choice between conflicting alternatives" (Shimahara 1970, p. 146).

In medical ethics education the need for a formal choice can be considered as absent, if proximally implicit. The activity may only involve, through engaging in dialogue with others, setting out and exploring various ethical positions and the way(s) in which they might be justified. Nevertheless this remains an enculturative experience and one could suppose that the articulation of a position implies at least the (imaginative) adoption of a particular ethical judgement in order to make the case, even whilst playing devil's advocate. Furthermore in medical ethics education there is also the adoption of the conceptual framework of modern medical ethics as the valid way of doing professional ethics. This indicates that such medical ethical enculturation is of plural concepts and the types of perspectives and judgements they lead to (their attendant logic and grammar). Medical ethics is not merely a number of singular concepts but an array or 'repertoire' of concepts and a particular logic suitable to the range of ethical issues one finds within medicine and to the culture of medicine as a subset of the wider cultural ecology more generally.

5.5.2 Enculturating Medical Ethical Metacognition

As indicated in the foregoing discussion one can hardly consider the enculturation of the concepts of medical ethics without slipping into a consideration of meta-cognition, the practice that forms the basis of their use. Whilst metacognitive ability might be thought of as a general cognitive skill it is in fact only ever articulated, or practiced, within a specific context through the use of some culturally specific, necessarily linguistic, concepts. Meaning is given through use and such use is indivisible from its logic or 'grammar' as use is constituted by such

logic or grammar. Consequentially if we understand concepts as being enculturated then it seems we must also understand metacognition as being enculturated.

If we recall that metacognition involves a "[s]kilful coordination of theory and
evidence" (Kuhn et al. 1988, p. 3) then in the context of medical ethics education
we might consider that case-based reasoning, which forms a major part of medical
ethics education, is one such activity. Case-based reasoning, whether in relation to
the 'thought experiment' or the 'real case', involves the presentation and framing
of the case or issue at hand (the presentation of evidence) alongside an analysis of
the ethically relevant features (introduction of theory or concepts). This initial
presentation is evaluated and some possible solutions are proposed or the kernel of
the ethical difficulty is identified. Very often this is then followed by a counterfactual varying of certain parameters of the case such that the nature of the ethical
problem can be elucidated through the use of theory to analyse and structure it.
However the ethical case and the ethical theory are in fact mutually illuminating,
particularly in the context of medical ethics education where the theory is being
acquired by medical students through its use. Furthermore, I would argue that this
is a kind of mental simulation or a 'rehearsal.' Various counterfactual possibilities
are considered and one's ethical responses are imagined. Even when conducted in
a solitary manner it nevertheless has a dialogical form as, at minimum, it is a
dialogue between case and theory whilst also being a kind of conversation with
oneself as various arguments and positions are considered, tested, accepted or
rejected. When reasoning the represented situation we use ethical theories and
concepts and so we are mentally rehearsing an ethical dialogue for the purposes of
co-ordinating theory and evidence. This can be considered a case of enculturating
metacognition.

5.5.3 Enculturating Beliefs[6]

I have focused on the primary function of medical ethics education: the teaching of
concepts and cognitive skills that form the basis of modern medical ethical
deliberation and reflection. However it can also be understood as a medium though
which the ethical beliefs of the profession are imparted to students and a venue that
provides students with the opportunity to develop their own perspectives in relation to the 'beliefs' of the profession. If they are to practice medicine medical
students must be made aware of the ethical precepts of the profession, as they are
required to adhere to them. Whilst medical professionals are independent practitioners they are also epistemologically and morally interchangeable. Patients can

---

[6] Belief here is not used in the loose sense of 'what people imagine, rather than know, to be true,'
something which usually refers to factually unsupported beliefs in Father Christmas, say. Here
belief refers, in the philosophical sense, to 'what people think or know' which includes factually
supported or logically reasoned knowledge such as biomedical or medical ethical knowing. In
philosophical terms knowledge is a *species* of belief.

expect the same diagnosis and moral perspective from all members of the same profession. Whilst there is room for valid differences in opinion or even dissent as, for example, in the case of conscientious objection to abortion, medical professionals must still discharge their responsibilities by, in this case, referring patients to a non-conscientiously objecting colleague in a timely fashion. Various professional bodies such as the GMC, the BMA, the Royal Colleges, and other specialist groups of professionals issue advice an guidance on the moral aspects of modern medical practice. Of course no medical professional is, or can be expected to be, completely cognisant of the range of guidelines or the detail of advice available. However in the course of medical education and, particularly, in the context of formal medical ethics education medical students are introduced to several of these guidelines and advice, particularly that issued by the GMC that demands compliance. In this way medical students become familiar with such advice (and where they might find it) and, more importantly the way in which this advice is structured which, of course, reflects the logic and grammar of modern medical ethics.

In the course of their medical education medical students develop their ethical beliefs. However, as I have argued, they do so as nascent professionals and so the beliefs developed are primarily produced for the purpose of becoming professionals. We can contrast this with the beliefs and arguments discussed in the philosophy classroom for the purpose of exploring the merit of ethical arguments on their own terms, the pursuit of ethical 'truth' and the development of the particular analytic skills native to philosophy. It should be no surprise that, particularly initially, the ethical opinions and beliefs developed by medical students accord with those of the profession even to the degree that research can conclude that medical students have 'regressed' in Kohlbergian terms. This is because medical students initial professional ethical development is accompanied by their acceptance of the (moral) authority of the medical profession and its institutions. Once medical students have developed an appreciation for the moral landscape and ethical discourses of medicine they have, in effect, also develop a familiarity for the way in which the profession (and, to a degree, the wider academic discipline of applied medical ethics) engages, structures and attempts to 'solve' those same problems. Enculturating medical students into the commonly accepted ethical beliefs that govern, structure and facilitate modern medical practice is an essential aspect of their medical training and fundamental to future medical practice.

5.5.4 Enculturating Thinking Dispositions

Thinking dispositions are acquired through the exercise of our reflective faculties and the thinking dispositions relevant to the practice of medical ethics are produced through encouraging medical students to exercise those reflective faculties as part of their medical ethics education. Through the enculturation of concepts, the meta-cognitive activity that is medical ethical reflection, and beliefs about the ethics of medicine the relevant thinking dispositions are produced. However, Shimahara's account of enculturation indicates that it involves conscious deliberation, reflection, choice and reasoning and, unlike concepts, meta-cognition (the activity of co-ordinating theory and evidence) and beliefs. Thinking dispositions are not available for the seemingly requisite direct and conscious examination. We might then consider thinking dispositions to be a second order product of formal ethical education; an unavoidable by-product of engaging in the relevant cognitive activity. Such a perspective echoes the way in which corporeal dispositions are produced through repeated physical activity as well as the sociological perspective that medical education necessarily transmits to medical students more than is stated in official medical curricula. In short this perspective reminds us that enculturation is not fully distinct from socialization. Rather they are two ends of the spectrum of social and cultural reproduction and both are aspects of Bourdieu's collective enterprise of inculcation. Entailed in this view is the fact that concepts (thinking tools), meta-cognition (the way of thinking that attends those tools), the beliefs they are used to (re)form and the thinking dispositions their use develops are intertwined. Enculturation offers a comprehensive picture of their collective development and one that compliments the well established perspective of the student's socialisation during their medical education and apprenticeship.

What this perspective offers is a specific account of the way in which the formal ethical thinking of modern medicine is transmitted to medical students that compliments the broader notion of reflection and reflective practice. During their education, training and apprenticeship medical students are encouraged to use and develop their reflective capacities both in specific senses, such as in medical ethics, but also more generally. Through reflection students develop a broad understanding of medicine and medical practice. Medial ethics education involves the enculturation of the more disciplined reflective practice that I have termed metacognition. This being the case we should understand reflection, a signature pedagogy of the professions (Shulman 2005), to be not only the *medium* of medical education but also, to a large degree, its *message*. The cognitive aspects of the medical apprenticeship constitute medical students as reflective medical professionals. Medical ethics education is one specific and disciplined aspect of the cognitive medical apprenticeship.

5.6 Conclusion

In this chapter I have introduced the notion of practice as central to a socio-cultural understanding of education and learning. Building on the view that thinking is a practice I proposed that concepts ought to be understood as tools for thinking with and that the mind should be perceived as a skilled organ. Furthermore I made clear the view that all practices, including teaching and learning, are situated. A practice that has been learnt in a particular context cannot necessarily be easily transferred to another situation. With regards medical ethics education I argued that neither the formal 'classroom' practice nor the informal 'clinical' practice should be seen as having primacy. Medical ethics is a practice that takes place in formal and informal context. It involves both reflection-in-action and reflection-on-action. Formal medical ethics education, and its enculturative results, is more directly linked to the latter practice.

References

Abend, G. 2012. What the science of morality doesn't say about morality. *Philosophy of the Social Sciences*. http://pos.sagepub.com/content/early/2012/08/16/0048393112440597.

Anspach, R.R. 1988. Notes on the sociology of medical discourse: The language of case presentation. *Journal of Health and Social Behavior* 29(4): 357–375.

Atweh, B., H. Forgasz, and B. Nebres. 2001. *Sociocultural research on mathematics education: An international perspective*. Mahwah: Lawrence Erlbaum Associates.

Bicknell, D.J. 1985. Current arrangements for teaching medical ethics to undergraduate medical students. *Journal of Medical Ethics* 11(1): 25–26.

Bloch, M. 2005. *Where Did Anthropology Go? Or: The Need For 'human Nature'*. In Bloch, M. Essays on Cultural Transmission. Oxford: Berg. 1–20.

Bourdieu, P. 1999. Scattered remarks. *European Journal of Social Theory* 2(3): 334–340.

Bourdieu, P. 2000. *Pascalian meditations*. Cambridge: Polity Press.

Brinkmann, S. 2010. *Psychology as a Moral Science: Perspectives on Normativity*. London: Springer.

Brockett, M., E.L. Geddes, M. Westmoreland, and P. Salvatori. 1997. Moral development or moral decline? A discussion of ethics education for the health care professions. *Medical Teacher* 19(4): 301–309.

Brosnan, C. 2008. The sociology of medical education: The struggle for legitimate knowledge in two English medical schools (Unpublished Ph.D. Thesis). University of Cambridge.

Brown, J.S., A. Collins, and P. Duguid. 1989. Situated cognition and the culture of learning. *Educational Researcher* 18(1): 32–42.

Brubaker, R. 1993. Social theory as habitus. In *Bourdieu: Critical perspectives*, ed. C. Calhoun, E. LiPuma, and M. Pstone, 212–234. Chicago: University of Chicago Press.

Chalmers, P., A. Dunngalvin, and G. Shorten. 2011. Reflective ability and moral reasoning in final year medical students: A semi-qualitative cohort study. *Medical Teacher* 33(5): e281–e289.

Cole, M. 1985. The zone of proximal development: Where culture and cognition create each other. In *Culture, communication, and cognition: Vygotskian perspectives*, ed. J.V. Wertsch, 146–161. New York: Cambridge University Press.

Dreyfus, H.L., and Dreyfus, S.E. 1990. What is moral maturity? A phenomenological account of the development of ethical expertise. In *Universalism* Vs. *Communitarianism*, ed. D. Rasmussen, 237–264. MIT Press, Cambridge.

Dreyfus, H.L., and S.E. Dreyfus. 1991. Towards a phenomenology of ethical expertise. *Human Studies* 14(4): 229–250.

Dreyfus, H.L., and S.E. Dreyfus. 2004. The ethical implications of the five-stage skill-acquisition model. *Bulletin of Science, Technology and Society* 24(3): 251–264.

Gellatly, A. (ed.). 1986. *The skilful mind: An introduction to cognitive psychology*. Great Britain: Open University Press.

Gerrans, P. 2005. Tacit knowledge, rule following and Pierre Bourdieu's philosophy of social science. *Anthropological Theory* 5(1): 53–74.

Gillett, G. R. 1987. Concepts, Structures, and Meanings. *Inquiry* 30(1): 101–112.

Gilligan, C. 1993. *In a different voice: Psychological theory and women's development.* Cambridge: Harvard University Press.

Goldie, J. 2000. Review of Ethics Curricula in Undergraduate Medical Education. *Medical Education* 34(2): 108–119.

Grenfell, M. 2009. Bourdieu, Language, and Literacy. *Reading Research Quarterly* 44(4): 438–448.

Guile, D., and M. Young. 1998. Apprenticeship as a conceptual basis for a social theory of learning. *Journal of Vocational Education and Training* 50(2): 173–193.

Hren, D., A. Vujaklija, R. Ivanišević, J. Knežević, M. Marušić, and A. Marušic. 2006. Students' moral reasoning, machiavellianism and socially desirable responding: Implications for teaching ethics and research integrity. *Medical Education* 40(3): 269–277.

Hren, D., M. Marušić, and A. Marušić. 2011. Regression of moral reasoning during medical education: Combined design study to evaluate the effect of clinical study years. *PLoS ONE* 6(3): e17406.

Kohlberg, L. 1981. *The philosophy of moral development: moral stages and the idea of justice,* vol. 1. New York: Harper-Collins.

Kuhn, D., E. Amsel, M. O'Loughlin, and H. Beilin. 1988. *The development of scientific thinking skills.* London: Academic Press.

Lave, J. 1988. *Cognition in practice: Mind, Mathematics and Culture in Everyday Life.* Cambridge: Cambridge University Press.

Lave, J. 1996. Teaching, as learning, in practice. *Mind, Culture, and Activity* 3(3): 149–164.

Lave, J., and E. Wenger. 1991. *Situated learning: Legitimate peripheral participation.* New York: Cambridge University Press.

Lin, C., K. Tsou, S. Cho, M. Hsieh, H. Wu, and C. Lin. 2012. Is medical students' moral orientation changeable after preclinical medical education? *Journal of Medical Ethics* 38(3): 168–173.

Lind, G. 2000. Moral regression in medical students and their learning environment. *Revista Brasileira de Educacao Médica* 24(3): 24–33.

Lipman, M. 2003. *Thinking in education* 2nd ed., Cambridge: Cambridge University Press.

Luke, H. 2003. *Medical education and sociology of medical habitus: It's not about the stethoscope!.* Dordrecht: Kluwer Academic Publishers.

MacIntyre, A. 1981. *After virtue: A study in moral theory.* London: Duckworth.

Mercer, N. 2000. *Words and minds: How we use language to think together.* Oxford: Routledge.

Miles, S.H. 1989. Medical ethics education: Coming of age. *Academic Medicine* 64(12): 705–714.

Patenaude, J., T. Niyonsenga, and D. Fafard. 2003. Changes in students' moral development during medical school: A cohort study. *Canadian Medical Association Journal* 168(7): 840–844.

Peters, R.S. 1963. Reason and habit: The paradox of moral education. In *Moral education in a changing society.* ed. Niblett, W.R. pp.47–48, London: Faber and Faber.

Piaget, J. 1997. The moral judgment of the child. Illinois: Free Press.

Reed, H.J., and J. Lave. 1979. Arithmetic as a tool for investigating relations between culture and cognition. *American Ethnologist* 6(3): 568–582.

Rogoff, B. 1992. *Apprenticeship in thinking: Cognitive development in social context.* New York: Oxford University Press.

Roth, W.M. 2001. Enculturation: Acquisition of conceptual blind spots and epistemological prejudices. *British Educational Research Journal* 27(1): 5–27.

Roth, W.M., and Y. Lee. 2007. Vygotsky's neglected legacy: Cultural-historical activity theory. *Review of Educational Research.* 77(2):186–232.

Schatzki, T.R., K.D. Knorr-Cetina, and E. Von Savigny (eds.). 2000. *The practice turn in contemporary theory.* Abingdon: Routledge.

Self, D.J., M. Olivarez, and D.C. Baldwin Jr. 1988. Clarifying the relationship of medical education and moral development. *Academic Medicine* 73(5): 517–520.

Sfard, A. 2007. *Commognition: Thinking as communicating, the case of mathematics*. New York: Cambridge University Press.

Sfard, A. 2010. *Thinking as communicating: Human development, the growth of discourses, and mathematizing*. New York: Cambridge University Press.

Shimahara, N. 1970. Enculturation-a reconsideration. *Current Anthropology* 11(2): 143–154.

Shulman, L.S. 2005. Signature pedagogies in the professions. *Daedalus* 134(3): 52–59.

Sinclair, S. 1997. *Making doctors: An institutional apprenticeship*. Oxford: Berg Publishers.

Splitter, Laurance J. 2010. Dispositions In Education: Nonentities Worth Talking About. *Educational Theory* 60(2): 203–230.

Sriraman, B., and L. English (eds.). 2010. *Theories of mathematics education*. Heidelberg: Springer.

Strauss, C. 1984. Beyond 'Formal' Versus 'Informal' Education: Uses of Psychological Theory in Anthropological Research. *Ethos* 12(3) (Autumn 1984): 195–222.

Strauss, C., and N. Quinn. 1998. *A cognitive theory of cultural meaning*. Cambridge: Cambridge University Press.

Stueber, K.R. 2005. How to think about rules and rule following. *Philosophy of the Social Sciences* 35(3): 307–323.

Talja, S. 2010. Jean Lave's practice theory. In *Critical theory for library and information science*, eds. G.J. Leckie, L.M. Given, and J. Buschman, 205–220, California: Libraries Unlimited (ABC-CLIO).

Taylor, C. 1985. Practice as social theory. In *Philosophy and the human sciences: Philosophical papers*, ed. C. Taylor, 91–115. Cambridge: Cambridge University Press.

Taylor, C. 1993. To follow a rule. In *Bourdieu: Critical perspectives*, eds., C. Calhoun, E. LiPuma, and M. Postone, 55–60. Chicago: University of Chicago Press.

Tishman, S., E. Jay, and D.N. Perkins. 1993. Teaching thinking dispositions: From transmission to enculturation. *Theory into Practice* 32(3): 147–153.

Vygotsky, L.S. 1981. The genesis of higher mental functions. In *The Concept of activity in soviet psychology*, ed. J. Wertsch, 144–188. Armonk: M. E. Sharpe.

Wenger, E. 1999. *Communities of Practice: Learning Meaning and Identity*. Cambirdge, UK: Cambridge University Press.

Wass, R., T. Harland, and A. Mercer. 2011. Scaffolding critical thinking in the zone of proximal development. *Higher Education Research and Development.* 30(3): 317–328.

Zigon, J. 2008. *Morality: An anthropological perspective*. Oxford: Berg.

Chapter 6
Conclusions

6.1 Introduction

In the preceding chapters I have discussed a range of theoretical perspectives on medical ethics education. However this has not been offered as a singular 'grand 'theory' of medical ethics education but, rather, it amounts to a number of over-lapping and interlinked theoretical lenses that can be used to illuminate the subject from a number of directions. I have given shape to the topic through two chapters focused on the history of medical ethics education in the UK. In these I focused on how the development of medical ethics education relates to changes that were contemporaneously occurring within medical education itself. This included the (global) development of the reflective paradigm as a signature pedagogy of medical and professional education (Shulman 2005) as well as the introduction 'outsiders' (or 'insider-outsiders') to UK medical education. Just as these have became embedded within medical culture so has the practice of medical ethics. This has modulated the critical function of medical ethics, a fact that is indicative of its success.

I have offered an interdisciplinary perspective on medical ethics education guided by an over-arching social theoretical perspective centered on professional reproduction. What connects these threads is their collective basis in practice and dispositionalism. Before offering some concluding remarks I address more directly the notion of dispositionalism and explore what it has to offer our understanding of medical ethics as an aspect of professional practice and as a part of medical education.

6.1.1 A Dispositional Thread

If the various perspectives I have discussed have a single thread that renders them cohesive then it is an underlying dispositionalism. However, as Schwitzgebel points out, dispositionalism is usually understood as a reductive project that aims

N. Emmerich, *Medical Ethics Education: An Interdisciplinary and Social Theoretical Perspective*, SpringerBriefs in Ethics, DOI: 10.1007/978-3-319-00485-3_6, © The Author(s) 2013

to show how "beliefs (in particular) or mental states (in general) can be trans-
formed or 'reduced' into talk about other, less objectionable things" (2002,
p. 258), primarily observable behaviour. In contrast his account of dispositionalism
encompasses the phenomenal and cognitive aspects of belief as well as their
behavioural correlates. His account can be understood as reintroducing 'the subject' to a dispositional theory of belief. As such it provides a theoretical space
within which thinking, reflection, reflexivity and subjectivity can be accommo-
dated alongside, rather than neglected in favour of, observable activity i.e.
behavior. This is, of course, an essential component of any social (scientific)
theory and stands in marked contrast to a dispositionalism that draws on the natural
sciences for inspiration. Schwitzgebel's position is interesting not only for its
phenomenal and cognitive dimension but also for the way in which he prefers to
consider its success in terms of its 'usefulness' rather than purely in terms of
metaphysical/ontological 'truth' and simplicity (2002, pp. 269–270). Thus his is a
pragmatic approach intended to light the way of others who are seeking to conduct
empirical research into belief in the social, cognitive and psychological sciences.
As such it accords with the interdisciplinary approach I have taken here.

Schwitzgebel claims his phenomenal, dispositional account of belief approach
is more successful than the standard accounts and accords with our intuitions
concerning belief and its attribution (2002, pp. 269–270). It is specifically constructed to accommodate not only our stereotypical beliefs -e.g. the belief that
tigers have four legs and are stripy or that the patient's autonomy is to be
respected—but also for various patterns of belief deviation (2002, pp. 263–266).
Of particular interest here is his account of Unreflective Inconsistency, Peripheral
Ignorance and Developing Beliefs.[1] He suggests the latter is closely related to the
previous two, the implication being that development of beliefs tends to mean
examining and eradicating peripheral ignorance through reflective processes. For
example, in the early part of their training a medical student may believe that
respecting the autonomy of the patient is a principle of medical ethics and that
some patients, children say, do not have the capacity for autonomy. However the
medical student may be (peripherally) ignorant of the way in which the transition
from childhood to adulthood, from the non-autonomous to the autonomous patient,
is handled in medical ethical thinking and practice. Through reflecting on the
beliefs that they have and being introduced to the problem and relevant cases, e.g.
the Gillick case, and exploring various ways of thinking about the problem, the
medical student can develop their beliefs about patient autonomy and the grey area
between the autonomous and non-autonomous patient. The substantive ground of
such belief development (the transition from childhood to adulthood) may inform,
but does not completely determine, other comparable grounds, such as issues
concerning the autonomy of patients who are developing dementia.

---

[1] In discussing the development of beliefs Schwitzgebel refers to Vygotsky and, implicitly, his
zone of proximal development as well as the idea of 'scaffolding' (2002, p. 265).

Adopting a dispositional approach to belief allows us to accommodate an important feature of human activity. It is often the case that we act in accordance with what we know but without making our beliefs fully 'occurrent' or bringing them into our conscious minds. Some of our beliefs are tacit and embodied. As such they are not amenable, possibly in principle or perhaps as a simple matter of contingent fact, to being brought to mind, at least not in a propositional form. Schwitzgebel cites the case of reversing a car into a driveway. The driver can accomplish the task but, if asked to explain precisely how it was done will, more often than not, be unable to do so with any degree of accuracy. Indeed if one concentrates on thinking about how to do certain embodied tasks it is often the case that simultaneously doing that task becomes more difficult. However, developing the ability to accomplish certain embodied tasks is achieved, at least in part, through thinking about them. Consider, for example, the way in which we learn to drive or the way in which a tennis player is coached. Language is part of the way in which our corporeal procedural know-how can be acquired and developed but it can become redundant. At least, until we engage in the practice of teaching someone else to do the same task.

We can apply this thinking to the ethical beliefs developed by medical students during their training. As I have discussed if an individual medical professional believes that abortion is morally wrong they do not need to rehearse the arguments when a patient consults them about an abortion, nor do they need to rehearse the arguments around referral. They simply comply with what it is that they already believe, something that may require only a minimal degree of conscious reflection. In such cases the relevant belief(s) are not necessarily phenomenally or cognitively occurrent, they can remain purely dispositional influences on action. In other cases there might be some phenomenal experiences. When dealing with a patient and explaining a diagnosis and treatment, a doctor will be implicitly providing information for the purposes of informed consent. If that patient reacts in a way that implies they have not properly understood, the ethical doctor will be disposed to think 'this patient has not understood, I need to find another way to explain myself.' On continuing the endevour the doctor may come to the conclusion that the patient lacks the capacity to give their informed consent and they will then be disposed to engage in further reflection on the implications of this and the best course of action that might subsequently be taken.

Like Schwitzgebel, Bourdieu's "conception of knowledge is, in effect, a dispositional one, which identifies knowledge [i.e. belief] with the socially acquired capacities, propensities and tendencies of an agent to act appropriately in given circumstances" (Gerrans 2005, p. 54). Schwitzgebel's expanded notion of dispositional belief (knowledge) as phenomenal indicates that acting appropriately may include having certain sorts of mental and emotional experiences. Furthermore, in the case of medical ethics, acting appropriately, and not simply re-acting appropriately, may require further reflection on the right course of action to take. On the basis of what they know, which we should take to include knowing-how and knowing-that, medical professionals can be disposed not only to certain forms of action, but also to certain phenomenal experiences and to undertake further ethical reflection and/or metacognition which, lest we forget, I have argued is a form of action. These forms of dispositional experiences are intertwined and reflexive, which is to say that on the basis of our actions, phenomenal experiences and thinking we become disposed to undertake further actions, have further phenomenal experience and to engage in further reflective activity. This does not see the return of the rational subject but, rather, is one way of construing a dispositional subject that has been constructed as such through the collective enterprise of inculcation, through socialisation and enculturation.

6.1.2 Beyond the New Synthesis in Moral Psychology

It could be tempting, in our neuro-maniacal times—an affliction to which (bio)ethics seems particularly susceptible (Tallis 2011)—to suggest that what it is to acquire a certain way of ethical thinking is to inscribe our brains with a particular neurological structure, to develop a set of pathways and synaptic connections that our neurons are inclined or 'disposed' to follow. However this would be to reintroduce an individualism of a potentially reductive and deterministic sort that Schwitzgebel's phenomenal dispositionalism and the various dispositional social theories discussed in this work are against. Whilst I have little doubt that there is some truth in the neurological perspective, I have every doubt that this is a useful way to think about things. Introducing such a perspective into this work risks reintroducing the asocial individual and tempts one to see ethics in equally individualist terms. This kind of individualism can be seen in the new synthesis in moral psychology. A perspective that raises a number of problems that can be responded to using the social perspective I have articulated.

Haidt suggests that our supposedly rational ethical reasoning is, in fact, largely determined by our affective emotional responses. In short, he thinks it is the emotional dog that wags its rational tale (Haidt 2001). Thus the work of Haidt and his interlocutors suggests that there are two cognitive processes involved in the formation of moral judgements. The first is fast, intuitive, and affective and the other is slow, reflective and deliberative (Haidt and Kesebir 2010, pp. 801–808). To be fair to Haidt and his interlocutors he does not think moral judgements are entirely dictated by our intuitive (emotional) responses and, furthermore, he suggests that 'moral thinking is for social doing' (Haidt and Kesebir 2010, p. 809). However others have suggested that our moral (e)valuations can shape our factual beliefs about ethical issues (Liu and Ditto 2012). Nevertheless, as Haidt's views on the psychology of moral development are unclear, the degree to which he thinks these two cognitive processes are interrelated is uncertain. Certainly they are intertwined at the phenomenological level. However given that Haidt suggests that there is variation in the "degree to which [cultures] construct, value, and teach virtues based on the five intuitive foundations"[2] (Haidt and Graham 2007, 104) then we can consider the content of each of these arenas of "affective reactions" to be the product of moral socialisation. However whilst the affective component is considered to have a genealogical heritage the same is not, at least not explicitly, stated of our deliberative moral or, more accurately, ethical responses.

However I have argued the moral socialization and the ethical enculturation of medical students are intertwined. In the course of their medical education medical students are inducted into a moral and ethical landscape that is both affectively and deliberatively structured. Using the tools of formal medical ethics they are encouraged to engage in ethical deliberation about the moral and ethical aspects of medical practice. In doing so medical students engage with their own emotional and affective responses such that they can be altered or put to one side when adopting the role of the professional. In doing so they *develop* both professionally and personally. The affective (emotional) and deliberative, reflective and meta-cognitive structures of medicine are interlinked. In this view the reasoning of medical ethics is not merely a function of individuals but also of a community of practice. The concepts and conclusions of medical ethical reflection structure the moral landscape of modern medical culture. Thus we can differentiate between socialization and enculturation, and correlate them with morality and ethics respectively. Nevertheless, both are part of the collective process of inculcation undergone by medical students during the medical apprenticeship.

In this view the ethical justifications produced by our slow, reflective and deliberative cognitive process can no longer be seen as, merely, a post hoc justification for our moral responses formed through the fast, intuitive and affective psychological reaction as they are intertwined. Haidt's view is rendered overly synchronic, as it is focused on particular instances of moral judgements. Adopting this perspective makes it appear that our affective moral reactions are in the driving seat. However, by adopting a broader diachronic perspective our ethical deliberations cannot be merely 'post-hoc' because they are not produced *de novo*. Our ethical concepts, beliefs and ways of thinking are dispositionally embodied and this embodiment is achieved through rehearsing, *practicing*, the formation of those beliefs prior to encountering those moral landscapes and ethical dilemmas that will become everyday or, in a sense, 'mundane.' This is the primary purpose of a formal ethics education delivered during the medical apprenticeship.

---

[2] These are: Harm/Care; Fairness/Reciprocity; Ingroup/Loyalty; Authority/Respect; Purity/Sanctity (Haidt and Graham Haidt 2007).

6.1.3 Professional Ethics as an Orthogonal Medical Specialism

The view I have presented here, more or less explicitly, involved a vision of medical ethics not only in medical practice but also as a practice of medicine. If we consider medicine as a social field we can appreciate that there are various kinds of ethical practice that occur in various aspects of its institutions, structures and cultures. The professional practice of ethics is not restricted to the clinic but also encompasses reflective moments away from but directly concerned with a specific set of circumstances, this includes the formal opportunity for reflection provided by the Clinical Ethics Committees. However some ethical practices of medicine do not directly concern a specific set of circumstances or a specific patient but consider a general set of circumstances populated by generalized patients and healthcare professionals. To a degree such generalization forms a part of the metacognitive practices of everyday reflection. However the medical profession also undertakes such activities *sui generis*. For example CECs are often asked to comment on policy and the formal structures of the profession, the GMC, The BMA, the Royal Colleges and other kinds of organisations and interest groups also offer their perspectives in the form of guidance, discussion papers, the organisation of conferences and working groups. We might also point to those medical professionals who contribute to the 'bioethical' literature. The JME, for example, often carries papers by medical professionals. All of these activities form an integral part of the ethical culture of modern medicine. Certainly not all of them can be undertaken following the basic introduction to medical ethics medical students usually receive in their undergraduate education.

However, in the same way that medical students undergraduate education in surgery, pediatrics and nephrology stands in relation to the practice of surgery, pediatrics and internal medicine so their medical ethics education stands in relation to these broader ethical practices of the profession. Echoing the 1966 sediments of ASME and the BJME with regard to medical education my view is that medical ethics is, or can be, a medical specialization. However it is one that, like medical education, is orthogonal to the 'true' medical specializations which have their basis in the anatomical and biomedical sciences. Expertise in medical ethics cuts across the biomedical disciplines of medical practice.[3] Certainly such expertise can be particularly acute when coupled with expertise in a branch of biomedicine but, nevertheless, once appropriately developed such expertise can taken into new contexts. Furthermore, whilst it does not make one a medical professional, non-clinically trained individuals can develop an expertise in medical or 'bio' ethics and make an important contribution to the ethical discourses of the profession.

---

[3] Although I disagree with the vast majority of commentators on the issue of (bio)ethical expertise and whilst I will not argue the point here I do believe that, if we accept and expand the perspective offered here, a cultural and democratically acceptable account of bio- and medical ethical expertise can be given.

Such expertise abuts and interacts with that of the profession and forms an important point of cultural exchange, an important point of external ethical input (and output), for the medical profession understood as a social field. Thus whilst a basic education in the ethics of medicine should be delivered to all medical students opportunity should be (and indeed is) provided to those who wish to pursue an interest either personally or professionally. To varying degrees and for varying ends and purposes medical professionals can and do specialize in medical ethics. In doing so they give shape to the boarder ethical aspects of the culture and structures of the medical profession; the very culture that medical students are introduced to and the very structures that the initial development of medical ethics education sought to change through its activities.

The metacognitive reasoning of individuals cannot be considered fully independent of the metacognitive reasoning of groups and the social structures in which it is embedded. They are fundamentally interrelated. Properly understood the ethical reasoning distributed across the medical profession provides a structural context within which the ethical (meta)cognition of individual medical professionals takes place. As a result the cultural practice of metacognition—in the specific form of medical ethics—is not and can never be truly monological or solely the property of any one individual as that individual is part of a larger community of practice. The internalization and widespread adoption of medical ethics on the part of the medical profession and medical professionals has resulted in it becoming an important structuring aspect of the field (embedded in ethical guidelines and advice) and a mundane or everyday part of professional practice. However we should not dichotomise these two forms but, rather, seek their connections. There are arenas where the practice of ethics goes beyond the mundane or everyday and there are spaces within (and without) medicine where professionals engage in critical ethical debate. For example there are formal spaces within professional institutions for ethical analysis (usually committees). As I have noted professionals also contribute to the bio- and medical ethical literature and to broader public, political and professional discourses about ethical issues. It is within these wider spheres that broader ethical questions are critically debated and, to the degree possible, settled. If the debates are to shape everyday medical (ethical) practice it is essential that clinicians are included at these levels. Just as it is essential that biomedical ethicists engage with and offer their insights to the medical profession it is essential that the medical profession has its own specialists in medical ethics that can, in return, engage with bioethics and thereby continue top propagate and develop modern medicines culture of medical and bio-ethics.

6.1.4 Implications for (Applied Philosophical) Bioethics

Shusterman, one of the leading philosophical interpreters of Bourdieu's social theory, has claimed that 'critical reflexivity' is not only "an explicit emblem of Bourdieu's thought, but also a traditional hallmark of philosophy itself"

(1999, p. 10). However, in the same edited collection, Pinto suggest that philosophy's "last great form of naivety" (1999, p. 109) is its weak understanding of social reflexivity, particularly in regards our (and its) linguistic practices. Whilst analytic reflexivity might be the *sine qua non* of philosophy this does not mean that contemporary philosophical culture does not have its reflexive blind spots. In my view applied philosophical ethics is possessed of this blind spot and, in the area of medical ethics education, I have sought to offer some corrective. This view does not undermine philosophical applied ethics per se. Rather, I have sought to reveal how medical ethics—an applied philosophical way of thinking – is taught, learned and, more or less implicitly, practiced. What this recommends is, I think, that we should recognise how bioethics reconstitutes the object(s) of its concern. Reubi (2012), for example, has demonstrated how the bioethical concept of informed consent configured research participants as reflective decision makers. The ethical subjectivity of medical professionals has been similarly configured.

Since Smith (2007, p. 76) argues that, recognized or not, a reflexive relationship between the subject and the object is what marks out the human sciences. One might argue that alongside history, sociology, (cognitive) anthropology and (cultural and social) psychology—the subjects discussed in the preceding chapters—(bio)ethics ought to recongise itself as a human science. As the example of Reubi demonstrates such a reflexive understanding can be found in many areas of bioethical research. However one might question whether applied ethics could do more to understand the reflexive relationship it has with those it seeks to influence and teach. Furthermore it could do more to recognize that the resolution of ethical dilemmas is not easy. In the literature there is often, at least in its presentation, an air of finality about the conclusion of ethical arguments. However Winch (1965) can be understood as suggesting if ethical dilemmas were amenable to final resolution they would not be ethical dilemmas.

Medical professional grapple repeatedly with not dissimilar ethical questions raised by cases. Whilst, as I have suggested, they develop particular ethical positions and respond accordingly and consistently this does not mean the ethical dilemma has been solved, resolved or dissolved. Or, as Talja has suggested of Lave's perspective: "It is the nature of dilemmas that they require managing contradictory principles and conflicting values. Dilemmas have no factual solutions or correct answers; resolving them is a matter of choosing between equally viable alternatives" (2010, p. 213). Despite the fact that the ethical difficulties of medical practice might receive consistent treatment and response this does not mean that have gone away. Rather the principles and concepts of medical ethics allow healthcare professionals to name, define, articulate and discuss ethical problems; to bring them out into the open and to factor them into the thick context of practice, diagnosis, treatment and illness. This perspective is consistent with the view that in the 'real-world' there is no formal distinction between 'problems' and their 'solutions' (Talja 2010, p. 212).

A reflexive realization of this on the part of applied philosophical bioethics would result in a reconfiguration of its relationship with those who are the subjects of its inquiries. It would result in a view that understands bioethics is embedded within and not determinate of healthcare practice. However philosophical applied ethics is marked by an "anti-ethos ethos ... [that] embeds the practice of disembedding" (Anderson 2005, p. 178). However, inescapably, it does have an ethos and it cannot escape the historical, social and cultural milieus in which it is embedded. Furthermore these milieus have involved embedding bioethics in other policy and practice contexts. The constitution of bioethics predicated on bringing academic perspectives on ethics into a relationship with aspects of our wider society. In doing so we create a concern for the practice/theory and practitioner/theorist relationship. Medical ethics education presents an attractive ground for the exploration of this relationship (cf. Cribb 2011). If we are to reflexively develop the practice of 'bioethics' then it must be orientated correctly. We require a proper understanding of the social and cultural locale in which it is embedded and ethos-full or thick approach to situated morality, ethics and individuals. As a disciplined practice academic applied philosophical (bio)ethics may be methodologically constituted as an a-cultural endeavor, however its relationship to medical practice, the field of healthcare and wider society it is, unavoidably, a cultural phenomenon. This perspective should be incorporated into the theoretical practices of applied bio-ethics.

6.1.5 Implication for Medical Ethics Education

The view I have offered here has not been designed to produce a new way of teaching medical ethics but, rather, to produce a broader understanding of how and why we currently teach medical ethics to medical students. On this basis we can, I think, renew calls for a greater integration of ethics across the medical curriculum, continue to pursue the medical humanities in undergraduate medicine and re-examine the still emerging notion of professionalism. Before examining each of these in turn it is worth echoing Kuhn's warning that "most educational programs designed to teach thinking skills have focused on teaching students about good thinking, rather than engaging them in the activity of thinking" (Kuhn 1991, p. 292). Medical ethics education has not fallen into the trap of solely teaching medical students *about* good ethical thinking and does engage them in the activity of ethical thinking. However in this distinction we can see something of the difference between the applied ethics classroom in philosophy and in medicine. It also recalls Irwin's (1987) concern that medical students should not over examine the foundations of morality lest the super-structure comes tumbling down. Medical students should be taught medical ethics and not necessarily the philosophy or meta-ethics of medical ethics.

This perspective also speaks to debates about whether we should teach philosophical moral theory to medical students (Lawlor 2007, 2008; Benatar 2007, 2009). It is likely to have little purpose unless the ethical discourse of contemporary medical culture happens to be conducted in such terms. This is unlikely to ever be the case. Furthermore the advantages of using 'pure' moral theory (and here we might note the absence of virtue ethics from most discussions of the place of moral theory in professional ethics education) is their supposed universal scope. Winch (1965) argues that regardless of any universal properties of a moral theory we must, in practice, abandon any vestige of hope for universality. In philosophical discourse ethical cases are constructed on the basis of what Benhabib (1985) calls 'generalized,' as opposed to 'concrete,' others. They are abstract individuals: 'A Doctor', not a specific doctor and 'A patient' not a specific person. However in practice it is these individual's and their appreciation of the moral situation, that construct the case. Winch proposes that any evaluation of a concrete moral situation must include an appreciation of the way in which its moral features strike the protagonists. In no small part this will be a function of their moral and ethical dispositions, their habitus, which Meisenhelder (2006) suggests fulfills the function of character in Bourdieu's social theory. Winch (1965, p. 169) argues that such a perspective renders the univeralizability principle idle. No matter how alike or comparable cases might seem they are essentially different as different concrete individuals with differing moral and ethical perceptions populate them.

If this is the case we might suppose that the effect of medical ethics education is to influence they way in which medical students, and healthcare professionals more generally, talk, and therefore think, about the moral and ethical parameters of professional practice. It is an attempt to get them talking, and thinking, in the same ethical language. Furthermore it is an attempt to develop medical student's moral and, particularly, professional ethical dispositions such that ethical issues raised in medical practice strike them in the same, or similar, manner. This does not permit the return of the universalizability thesis as Winch's challenge is not dismissed by the similarity of individual professional's medical ethical dispositions. The universalizability thesis relies on abstract individuals being theoretical interchangeable. Whilst medical professionals are considered to be epistemically and morally interchangeable in a practical sense no amount of training can make such concrete individuals truly, or theoretically, interchangeable. Attempting to render the moral and ethical dispositions, perspectives and character of medical professionals similar does not produce complete homogeneity but, rather, a certain degree of practical homogeneity. Thus the development of professional (moral and ethical) character on the part of individuals is central to the profession's moral coherence. Something that is essential to its sociological viability.

In the preceding paragraphs I have talked of the moral and ethical dispositions and character of medical professionals. Whilst in this book I have focused on discussing the ethical education or enculturation of medical students this has been clearly positioned in relation to their moral socialization. Whilst in philosophy 'morality' and 'ethics' are usually understood to be synonymous here I have treated the former as having greater emotional or affective content whilst the latter is more cognitive and conceptual. Nevertheless I have stressed their interconnection both developmentally and theoretically. I have also suggested, in Chap. 5, that the division I have drawn echoes a perceived need to integrate 'classroom' and 'clinical' medical ethics education. The view I have presented allows us to consider this integration in terms of connecting the ethical enculturation of medical students with their moral socialization.

This could be taken to mean that there should be greater opportunity for medical students to exercise their developing metacognitive abilities in the clinical context. Individual clinical leads could either facilitate this, as now takes place, or we might draw in the increasingly prevalent hospital Clinical Ethics Committee (CEC). Others have suggested that the CEC has an educational role (Larcher 2009, 48) and this could be extended to medical students. Simply observing a number of meetings is likely to be instructive and, perhaps with the assistance of those with CEC experience, medical students could be encouraged to form their own ethics committee for the purposes of discussing the issues they face. Medical ethics has been disseminated throughout the medical profession through the education of medical students. A similar strategy could be adopted with regard CECs as, whilst they are increasingly prevalent, they are, as yet, under utilized. Regardless of this potential wider benefit, through providing medical students with opportunities to connect formal medical ethical debate with their experiences in the clinical context, a greater connection between their enculturation and socialization can be fostered.

Another way in which this connection can be fostered is through a continued development of the humanities in the undergraduate medical curriculum. Whilst the history of medicine predates medical ethics in the medical curriculum it is nevertheless the case that the teaching of the medical humanities has experienced some growth over the past decade or so. It is usually claimed that teaching medical students the medical humanities will assist in their personal and professional development, make them better, more sensitive doctors and encourage them to perceive the wider socio-cultural context in which medicine exists. Whilst one might assume the medical humanities to be history, literature and the arts (film, drama, painting etc.) the social sciences are often included, particularly those that can shed light on the thick, socio-cultural contexts in which patients experience their illnesses and encounter medicine. Furthermore it is certain, although often implicitly so, that the philosophy of medicine is a humanities subject. As such we might and perhaps should (re)construe and expand medical and bio- ethics as a humanities subject (Emmerich 2011a). In rethinking the nature of medical ethics in practice and education I have been trying to draw attention to something Downie and MacNaughton also seek to illuminate. They suggest that "the diagnosis of a patient's illness and the analysis of an ethical problem have this in common: each is more like the interpretation of a difficult text than like either the scientific analysis of urine or the logical analysis of an argument" (Downie and Mac-naughton 2007, p. 135). Applied philosophical bioethicists logically analyse medical ethical arguments. The real challenge facing medical professionals is in interpreting the difficult moral and ethical 'text' of a particular patient or case. Certainly coming up with a (logical) argument can often be important but this is secondary to the broader challenge of remaining a humanistic practitioner in an age of biotechnological achievement.

Another recent development in medicine and medical education has been the advent of 'professionalism.' Given the sociological connection between morality and a profession and between ethics and being a professional, it is unsurprising to find the term being discussed in relation to other professions where it also has a strong moral and ethical component (e.g. Carr 2000). Considering this recent turn to professionalism or, particularly in the UK where it is linked to developments in the GMC including those I have discussed such as Tomorrow's Doctors (1993), this *renewed* medical professionalism (Irvine 1999, 2001) sheds light on some of the implication of our discussion. Whilst being concerned with ethics, and the right thing to do, the new professionalism is also concerned with being the right sort of doctor. Certainly commentators such as Christmas and Milward are resistant to the idea that there is any one vision of the good doctor suggesting the term "may now describe a network of 'family resemblances' between roles in different contexts" (2011, p. 7) a view that finds some empirical support (Spyridonidis and Calnan 2011). However whilst there may no longer be a singular vision of what it is to be a good doctor, there remains a strong sense in which being a good doctor is linked to being a certain sort of person. The view I have outlined above suggests that professional moral and ethical development is likely to impact on the personal development of medical students. Within the professionalism paradigm this link is clearer and, furthermore, it places greater constraints on the personal character- istics of medical doctors precisely because it is less explicitly prescriptive. In this one can see the echo of the GMC's historical reluctance to issue formal guidance on acceptable medical practice and instead rely on conceptions of gentlemanly conduct and the accumulation of precedent (Smith 1993).[4]

The challenge facing the new professionalism and, indeed, any socially and culturally institutionalised profession is to remain open to dissident professionals. Whilst one might disagree with, say, radical or 'anti' psychiatrists such as Thomas Szasz or R.D. Laing one can recognise that psychiatry would be the poorer if they, or their criticisms, were dismissed for a supposed lack of professionalism. Whilst this is an extreme example current concerns around the use of social media by healthcare professionals demonstrate precisely how far this new professionalism might intrude on the personal lives of medical professionals. There must be limits on the ways in which the new professionalism can act to restrict the personal freedom of individuals. We may legitimately demand that our healthcare profes- sionals act ethically but we cannot expect them to be saints. Furthermore what it is to act ethically is not easily determined and it may be that it often involves being open to ethical criticism and debate. Often such debate occurs at a social, or structural, rather than individual level and it is the profession's responsibility to engage with the contemporary moral and ethical debates that arise both internally and externally. As such medical education is, therefore, required to produce pro- fessionals that are prepared to engage in such debates and with medical

---

[4] This approach remained until the 1963 when the GMC published the first of its famous 'Blue Books' and, even then, can be considered in force until the 1976 revision.

interlocutors who range from patients and members of the public to politicians, other healthcare professionals and bioethicists from a variety of disciplinary backgrounds.

6.2 Conclusion

The social theoretical view of medical ethics education offered here was developed in my PhD (Emmerich 2011b). The ends I had in mind when doing so were multiple. Certainly, as I have presented here, I sought to offer a broader understanding of medical ethics education than can currently be found in the literature. However I also sought to offer a basis on which further empirical social scientific research could be conducted. These two aims are not fully distinct. Those who have read this book have, no doubt, done so because they are involved in medical ethics and medical ethics education. In reading this book I hope that they—you— have drawn on their own experience in assessing validity and usefulness of the perspective I offer. However, I also hope that they—you—have been prompted to reassess your experiences and that you have come to a new understanding of medical ethics education and what it is you do when teaching medical ethics. Empirical social scientific research is, simply, a more disciplined attempt to produce this kind of understanding and reassessment. I hope that others will take up my tools and further explore medical ethics education in a more empirical manner. However I do not wish to offer the unsatisfying conclusion that 'more research is needed.' What is needed is not only further empirical research but also an more involved dialogical engagement amongst those with an interest in medical ethics and medical ethics education.

Despite the interdisciplinary foundation and hopes of bioethics there remains too much misunderstanding entrenched in largely disciplinary miscommunication. As discussed Abend (2012) has suggested that moral psychology suffers from its focus on judgement and one could level a similar charge at bio- and particularly applied philosophical ethics. There is a lack of appreciation for the ethical and moral landscapes in which others exist and in which their 'judgements' are formed. In order that all concerned might uncover the lessons we have to teach each other about morality and ethics we must overcome these differences. All too often we are only concerned with what we can tell other people rather than in what we can learn from them and how they might offer us new insights into areas we consider our own. Whether they stem from the research I reference or in my experiences in teaching, learning and practicing (bio)ethics virtually all of the insights offered here have been drawn from my engagement with others. This engagement has involved agreement and disagreement, acceptance and rejection but also dialogue and mutual understanding. Nevertheless we must embrace, even if only at the simply level of comprehension, the perspectives of others. To become truly interdisciplinary bioethics must develop what Collins and Evans' (2007) have called interactional expertise. We must learn to be more co-operative with scholars whose discipline are not our own, with professionals whose practices are not our own and to engage with each other across our disciplinary, professional and other cultural contexts. After all, what else might the practice of morality and ethics be?

References

Abend, G. 2012. What the science of morality doesn't say about morality. *Philosophy of the Social Sciences*. http://pos.sagepub.com/content/early/2012/08/16/0048393112440597.

Anderson, A. 2005. *The way we argue now: A study in the cultures of theory*. Priceton: Princeton University Press.

Benatar, D. 2007. Moral theories may have some role in teaching applied ethics. *Journal of Medical Ethics* 33(11): 671–672.

Benatar, D. 2009. Teaching moral theories is an option: reply to Rob Lawlor. *Journal of Medical Ethics* 35(6): 395–396.

Benhabib, S. 1985. The generalized and the concrete other: The Kohlberg-Gilligan controversy and feminist theory. *PRAXIS International* 5(4): 402–424.

Carr, D. 2000. *Professionalism and ethics in teaching*. New York: Routledge.

Christmas, S., and L. Milward. 2011. *New medical professionalism: A scoping report for the health foundation*. London: The Health Foundation. http://www.health.org.uk/public/cms/75/76/313/2733/New%20medical%20professionalism.pdf?realName=JOGEKF.pdf (Accessed September 2012)

Collins, H.M., and R. Evans. 2007. *Rethinking expertise*. Chicago: University of Chicago Press.

Cribb, A. 2011. Beyond the classroom wall: Theorist-practitioner relationships and extra-mural ethics. *Ethical Theory and Moral Practice* 14(4): 383–396.

Downie, R., and J. Macnaughton. 2007. *Bioethics and the humanities: Attitudes and perceptions*. London: Routledge-Cavendish.

Emmerich, N. 2011a. Literature, history and the humanization of bioethics. *Bioethics* 25(1): 112–118.

Emmerich, N. 2011. Taking education seriously: Developing Bourdieuan social theory in the context of teaching and learning medical ethics in the UK undergraduate medical degree. Unpublished PhD Thesis, Queen's University, Belfast.

Gerrans, P. 2005. Tacit knowledge, rule following and Pierre Bourdieu's philosophy of social science. *Anthropological Theory* 5(1): 53–74.

Haidt, J. 2001. The emotional dog and its rational tail: A social intuitionist approach to moral judgment. *Psychological Review* 108(4): 814–834.

Haidt, J. 2007. The new synthesis in moral psychology. *Science* 316(5827): 998–1002.

Haidt, J. Graham, J. 2007. When Morality Opposes Justice: Conservatives Have Moral Intuitions That Liberals May Not Recognize. *Social Justice Research* 20(1): 98–116.

Haidt, J., and S. Kesebir. 2010. Morality. In *Handbook of social psychology*, vol. 2, ed. S.T. Fiske, D.T. Gilbert, and G. Lindzey, 797–832. Hoboken: Wiley.

Irvine, D. 1999. The performance of doctors: The new professionalism. *The Lancet* 353(9159): 1174–1177.

Irvine, D. 2001. Doctors in the UK: Their new professionalism and its regulatory framework. *The Lancet* 358(9295): 1807–1810.

Irwin, W.G. 1987. Medical ethics (presidential address to the Ulster Medical Society). *The Ulster Medical Journal* 56(1): 1–12.

Kuhn, D. 1991. *The skills of argument*. Cambridge: Cambridge University Press.

Larcher, V. 2009. The development and function of clinical ethics committees (CECs) in the United Kingdom. *Diametros* 22: 47–63.

Lawlor, R. 2007. Moral theories in teaching applied ethics. *Journal of Medical Ethics* 33(6): 370–372.

Lawlor, R. 2008. Against moral theories: reply to Benatar. *Journal of Medical Ethics* 34(11): 826–828.

Liu, B.S., and P.H. Ditto. 2012. What dilemma? Moral evaluation shapes factual belief. *Social Psychological and Personality Science.* http://spp.sagepub.com/content/early/2012/08/13/1948550612456045.abstract.

Meisenhelder, T. 2006. From character to habitus in sociology. *The Social Science Journal* 43(1): 55–66.

Pinto, L. 1999. Theory in practice. In *Bourdieu: A critical reader*, ed. R. Shusterman, 94–112. London: Blackwell.

Reubi, D. 2012. The human capacity to reflect and decide: Bioethics and the reconfiguration of the research subject in the British biomedical sciences. *Social Studies of Science* 42(3): 348–368.

Schwitzgebel, E. 2002. A phenomenal, dispositional account of belief. *Noûs* 36(2): 249–275.

Shulman, L.S. 2005. Signature pedagogies in the professions. *Daedalus* 134(3): 52–59.

Shusterman, R. 1999. Introduction: Bourdieu as philosopher. In *Bourdieu: A critical reader*, ed. R. Shusterman, 1–13. London: Blackwell.

Smith, R. 2007. *Being human: Historical knowledge and the creation of human nature.* Manchester: Manchester University Press.

Smith, R.G. 1993. The development of ethical guidance for medical practitioners by the general medical council. *Medical History* 37(1): 56–67.

Spyridonidis, D., and M. Calnan. 2011. Are new forms of professionalism emerging in medicine? The case of the implementation of NICE guidelines. *Health Sociology Review* 20(4): 394–409.

Talja, S. 2010. Jean Lave's practice theory. In *Critical theory for library and information science*, eds. G.J. Leckie, L.M. Given, and J. Buschman, 205–220, California: Libraries Unlimited (ABC-CLIO).

Tallis, R. 2011. *Aping mankind: Neuromania, Darwinitis and the misrepresentation of humanity.* Durham: Acumen Publishing.

Winch, P. 1965. Universalizability of moral judgements. *The Monist* 49(2): 196–214.

General Medical Council Publications

GMC, Education Committee. 1993 (Revised 2003, 2009). Tomorrow's Doctors: Recommendations on Undergraduate Medical Education. London, UK. www.gmcuk.org/Tomorrows_Doctors_1993.pdf_25397206.pdf (1993) and http://www.gmc-uk.org/education/undergraduate/tomorrows_doctors.asp (Accessed July 1, 2011).

Printed by Printforce, the Netherlands